The Art of Smooth Pasting

FUNDAMENTALS OF PURE AND APPLIED ECONOMICS

ADVISORY BOARD

K. ARROW, Stanford, California, USA
W. BAUMOL, Princeton, New Jersey, USA
W. A. LEWIS, Princeton, New Jersey, USA
S. TSURU, Tokyo, Japan

Fundamentals of Pure and Applied Economics is an international series of titles divided by discipline into sections. A list of sections and their editors and of published titles may be found at the back of this volume.

The Art of Smooth Pasting

Avinash Dixit
Princeton University, New Jersey, USA

A volume in the Stochastic Methods in Economic Analysis section

London and New York

Copyright © 1993 The Gordon and Breach Publishing Group.

All rights reserved.

First published 1993
Second printing 1994
Third printing 1999

No part of this book may be reproduced or utilized in any form or by any means, electronic or mechanical, including photocopying and recording, or by any information storage or retrieval system, without permission in writing from the publisher.

Printed and bound in Great Britain by
LSL Press Limited
4 Caxton Park
Bedford MK41 0TY

Transferred to Digital Printing 2003

Routledge is an imprint of the Taylor & Francis Group

Library of Congress Cataloging-in-Publication Data

Dixit, Avinash K.
    The art of smooth pasting / Avinash Dixit.
      p. cm.—(Fundamentals of pure and applied economics; v. 55)
    Includes bibliographical references and index.
    ISBN 3-7186-5384-2
    1. Economics, Mathematical. 2. Brownian motion processes. I. Title. II. Series.
HB135.D577   1993
330'.01'51—dc20   92-45249
                                                                             CIP

Contents

Introduction to the Series	vii
Preface	ix
1. Brownian Motion	1
1.1. Random Walk Representation	2
1.2. Itô's Lemma	4
1.3 Geometric Brownian Motion	6
1.4 Some Generalizations	8
2. Discounted Present Values	9
2.1. Present Values for Exponential and Polynomials	10
2.2. Present Values for Powers of Geometric Brownian Motion	13
2.3. A Basic Differential Equation for Present Value	14
2.4. Derivation by Discrete Approximation	15
2.5. The General Solution	16
2.6. Differential Equation for Geometric Brownian Motion	19
2.7 General Diffusion Processes	21
3. Barriers	22
3.1. The Basic Differential Equation	23
3.2. Geometric Brownian Motion	25
3.3. Stopping	25
3.4. Resetting	26
3.5. Reflection	26
3.6. Example: Price Ceiling	27
3.7. Example: Exchange Rate Target Zones	28
3.8 Transitional Boundary	30
3.9 Example: Temporary Suspension	31
4. Optimal Control and Regulation	32
4.1. Stopping	34
4.2. Example: Irreversible Investment	37

4.3.	Convex Costs: Continuous Control	38
4.4.	Lump-Sum Costs: Impulse Control	39
4.5.	Example: Menu Costs	41
4.6.	Linear Costs: Barrier Control	42
4.7.	Some Geometry and Intuition	43
4.8.	Example: Competitive Industry	45

5. Generalizations — 47
 5.1 Mean-Reverting Processes — 47
 5.2 Finite Horizon — 49

6. Some Characterization of Optimal Paths — 51
 6.1 Short Run: Time Until First Action — 52
 6.2 Long Run: Stationary Distribution and Average Action — 58
 6.3 Dynamics of Brownian motion: Kolmogorov Equations — 63

References — 69
Index — 71

Introduction to the Series

Drawing on a personal network, an economist can still relatively easily stay well informed in the narrow field in which he works, but to keep up with the development of economics as a whole is a much more formidable challenge. Economists are confronted with difficulties associated with the rapid development of their discipline. There is a risk of 'balkanization' in economics, which may not be favorable to its development.

Fundamentals of Pure and Applied Economics has been created to meet this problem. The discipline of economics has been subdivided into sections (listed at the back of this volume). These sections comprise short books, each surveying the state of the art in a given area.

Each book starts with the basic elements and goes as far as the most advanced results. Each should be useful to professors needing material for lectures, to graduate students looking for a global view of a particular subject, to professional economists wishing to keep up with the development of their science, and to researchers seeking convenient information on questions that incidentally appear in their work.

Each book is thus a presentation of the state of the art in a particular field rather than a step-by-step analysis of the development of the literature. Each is a high-level presentation but accessible to anyone with a solid background in economics, whether engaged in business, government, international organizations, teaching, or research in related fields.

Three aspects of *Fundamentals of Pure and Applied Economics* should be emphasized:

— First, the project covers the whole field of economics, not only theoretical or mathematical economics.
— Second, the project is open-ended and the number of books is

not predetermined. If new and interesting areas appear, they will generate additional books.

— Last, all the books making up each section will later be grouped to constitute one or several volumes of an Encyclopedia of Economics.

The editors of the sections are outstanding economists who have selected as authors for the series some of the finest specialists in the world.

Preface

Many recent stochastic dynamic models in economics and finance are based on the theory of Brownian motion and its control or regulation. A fully rigorous mathematical treatment of these topics, as found in most treatises on probability theory and stochastic processes, is beyond most economists' reach. That has deterred wider understanding and use of such models in economic research. The well-known exposition by Malliaris and Brock (1982) has met the needs of many financial economists, who need these techniques constantly and have greater reason to make a major investment in learning them. But other fields like international economics, macroeconomics, and even labor economics, have begun to find occasions to use Brownian motion and its control. Researchers in these fields have less reason to learn this theory with any rigor and detail, and would welcome an even simpler treatment. These notes attempt a very heuristic exposition aimed at such readers. Brownian motion is introduced as the limit of a random walk, and many of its important properties are derived as the limits of the corresponding results on random walks that can be proved using quite simple algebra. This approach has been used in books on stochastic processes, for example Cox and Miller (1965) and Bhattacharya and Waymire (1989); it was introduced in option pricing theory by Cox and Ross (1976) and Cox, Ross and Rubinstein (1979). Here I extend it to treat problems of control of Brownian motion, and derive the appropriate conditions by similar elementary algebraic methods followed by passage to limits.

The following topics are covered: (1) Itô's Lemma and the Kolmogorov equations; (2) calculation of expected present values of functions of Brownian motion, including cases when the process has absorbing or reflecting barriers; (3) optimal control and regulation of Brownian motion, including derivation of smooth pasting conditions for discrete stopping and resetting problems, and super contact conditions for barrier control problems; (4) first passage times and ergodic distributions, and their implications for the frequency and the amount of control in short and long runs.

No attempt is made at mathematical rigor or completeness of treatment. I use heuristic devices, and sketch limiting arguments without proof. Nor do I develop other important aspects of the underlying stochastic processes, such as Markov and martingale properties, except where they matter for my focus, namely certain problems of optimal control. My applications are drawn mostly from recent research in macroeconomics and international economics; financial applications are of longer standing and are well treated in many books and surveys.

For economists who want to go beyond what is offered here, the first stop should be Malliaris and Brock (1982). After that, one must turn to more advanced mathematics texts. Roughly in increasing order of difficulty and rigor, these include Cox and Miller (1965), Feller (1968, 1970), Karlin and Taylor (1975, 1981), Harrison (1985), Bhattacharya and Waymire (1989), Fleming and Rishel (1975) and Karatzas and Shreve (1988). Some specific references to these will be given in the text.

I am very grateful to numerous readers of the earlier version, particularly Alfredo Cuevas, Lars Svensson, William Brock and Marco Bonomo, for their comments, suggestions and corrections of errors both typographical and substantive. I also thank the National Science Foundation and the Guggenheim Foundation for financial support. This revised version was substantially completed during a stay at the London School of Economics, and I thank their Suntory-Toyota Centre for hospitality and support.

Avinash Dixit

The Art of Smooth Pasting

AVINASH DIXIT
Princeton University, New Jersey, USA

1. BROWNIAN MOTION

Brownian motion is a continuous-time scalar stochastic process such that, given the initial value x_0 at time $t = 0$, the random variable x_t for any $t > 0$ is normally distributed with mean $(x_0 + \mu t)$ and variance $(\sigma^2 t)$. The parameter μ measures the trend, and σ the volatility, of the process. This process was first formulated to represent the motion of small particles suspended in a liquid. We shall sometimes refer to a 'particle' performing the Brownian motion, x_t as its 'position', and a graph of x_t against t as its 'path'.

We can think of Brownian motion as the cumulation of independent identically normally distributed increments, the infinitesimal random increment dx over the infinitesimal time dt having mean μdt and variance $\sigma^2 dt$. Just as we would write a general normal (μ, σ) variable as $\mu + \sigma w$ where w is a standard normal variable of zero mean and unit variance, we can write

$$dx = \mu dt + \sigma dw, \tag{1.1}$$

where w is a standardized Brownian motion (Wiener process) whose increment dw has zero mean and variance dt. This is the usual shorthand notation for Brownian motion.

The (Itô) calculus of such infinitesimal random variables differs in some important ways from the usual non-random calculus. A fully rigorous treatment of Itô calculus is quite difficult. Therefore I shall develop a non-rigorous exposition that suffices for many economic applications. I shall approximate Brownian motion by a discrete random walk. Then the normal distribution arises as the limit of a sum of independent binary variables Δx over discrete time intervals Δt, when these go to zero in a particular way.

1.1. Random walk representation

Divide time into discrete periods of length Δt, and let space consist of discrete points along a line, Δh being the step-length or the distance between successive points. Let Δx be a random variable that follows a random walk: in one time period it moves up one step in space with probability p, and one step down with probability $q = 1 - p$. Note: Δh is a given positive number, and Δx is a random variable that takes values $\pm \Delta h$. Figure 1.1 shows all this compactly. We see various possible paths, with time marching downward and position shown horizontally. At each point in time and space, the probability of reaching it is also shown.

The mean of Δx is

$$E[\Delta x] = p\Delta h + q(-\Delta h) = (p - q)\Delta h. \tag{1.2}$$

Also,

$$E[(\Delta x)^2] = p(\Delta h)^2 + q(-\Delta h)^2 = (\Delta h)^2,$$

so the variance of Δx is

$$\begin{aligned} Var[\Delta x] &= E[(\Delta x)^2] - (E[\Delta x])^2 \\ &= [1 - (p - q)^2](\Delta h)^2 = 4pq(\Delta h)^2. \end{aligned} \tag{1.3}$$

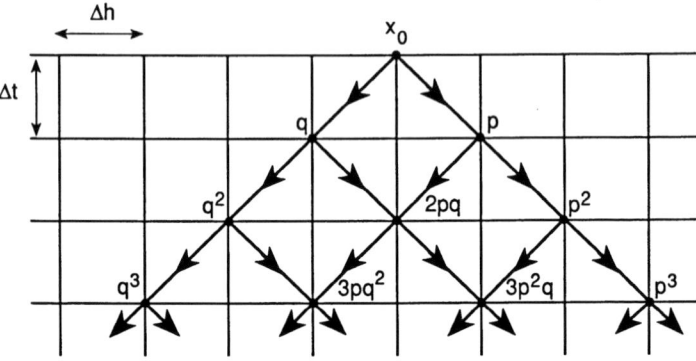

FIGURE 1.1 Random walk representation.

A time interval of length t has $n = t/\Delta t$ such discrete steps. Since the successive steps of the random walk are independent, the cumulated change $(x_t - x_0)$ is a binomial variate with mean

$$n(p - q)\Delta h = t(p - q)\Delta h/\Delta t$$

and variance

$$4npq(\Delta h)^2 = 4tpq(\Delta h)^2/\Delta t.$$

These expressions are minor modifications of the very familiar and elementary binomial distribution. There a 'success' in any one trial counts as 1 and occurs with probability p, while a failure counts as 0 and occurs with probability $q = 1 - p$. The (random) number of successes in n independent trials has expectation np and variance npq; see Feller (1968, pp. 223, 228). The expressions just above are perfectly analogous. Now success counts as Δh and failure as $-\Delta h$; therefore for example the variance is $4(\Delta h)^2$ times that of the usual binomial expression.

Now set

$$\Delta h = \sigma\sqrt{\Delta t}, \qquad (1.4)$$

and

$$p = \tfrac{1}{2}\left[1 + \frac{\mu}{\sigma}\sqrt{\Delta t}\right], \qquad q = \tfrac{1}{2}\left[1 - \frac{\mu}{\sigma}\sqrt{\Delta t}\right]. \qquad (1.5)$$

or

$$p = \tfrac{1}{2}\left[1 + \frac{\mu}{\sigma^2}\Delta h\right], \qquad q = \tfrac{1}{2}\left[1 - \frac{\mu}{\sigma^2}\Delta h\right]. \qquad (1.5')$$

Then

$$4pq = 1 - \left(\frac{\mu}{\sigma}\right)^2 \Delta t.$$

Substitute these into the above expressions, and let Δt go to zero. For given t, the number of steps goes to infinity. Then the binomial distribution converges to the normal, with mean

$$t\frac{\mu}{\sigma^2}\Delta h \frac{\Delta h}{\Delta t} = \mu t$$

and variance

$$t\left[1 - \left(\frac{\mu}{\sigma}\right)^2 \Delta t\right] \frac{\sigma^2 \Delta t}{\Delta t} \to \sigma^2 t.$$

These are exactly the values we need for Brownian motion. Thus we can regard Brownian motion as the limit of the random walk, when the time interval and the space step-length go to zero together, while preserving the relation (1.4) between them.

The mean of $(x_t - x_0)$ is μt, and its standard deviation is $\sigma\sqrt{t}$. For large t, we have $\sqrt{t} \ll t$; in the long run, the trend is the dominant determinant of Brownian motion. But for small t, we have $t \ll \sqrt{t}$, so volatility dominates in the short run.

Another manifestation of this volatility is seen by calculating the expected length of a path. We have

$$E(|\Delta x|) = \Delta h,$$

so the total expected length of the path over the time interval from 0 to t is

$$t\Delta h/\Delta t = t\sigma/\sqrt{\Delta t} \to \infty$$

as Δt goes to zero. For small but finite Δt, the total length of almost all sample paths is very large. Therefore each path must have many ups and downs and look very jagged. Most such sample paths are not differentiable. When discussing the expected rate of change, therefore, we must write $E[dx]/dt$, not $E[dx/dt]$.

The above derivation of Brownian motion as the limit of a random walk closely follows Cox and Miller (1965, Sections 2.2 and 5.2). A similar more recent exposition is in Bhattacharya and Waymire (1989, Section 1.7). More advanced but more direct definitions and analyses can be found in most books on stochastic processes, e.g. Karlin and Taylor (1975, chapter 7). A terse but quite rigorous exposition is in Harrison (1985, pp. 1–6).

1.2. Itô's Lemma

Suppose x follows Brownian motion with parameters (μ, σ). Consider a stochastic process y that is related to x by $y = f(x)$ where f is a given non-random function. We want to relate changes in y to those in x. The rules of conventional calculus suggest writing $dy = f'(x) dx$ and taking expectations. But this turns out to be wrong. Starting at $y_0 = f(x_0)$, consider the position a small amount of time t later.

$$y_t - y_0 = f'(x_0)(x_t - x_0) + \tfrac{1}{2}f''(x_0)(x_t - x_0)^2 + \ldots$$

Hence

$$\begin{aligned}E[y_t - y_0] &= f'(x_0)E[x_t - x_0] + \tfrac{1}{2}f''(x_0)E[(x_t - x_0)^2] + \ldots \\ &= f'(x_0)\mu t + \tfrac{1}{2}f''(x_0)[\sigma^2 t + \mu^2 t^2] + \ldots \\ &= [\mu f'(x_0) + \tfrac{1}{2}\sigma^2 f''(x_0)]t + \ldots,\end{aligned}$$

where in each case the dots represent terms in higher powers of t that can be ignored when t is small. But note that the second order term in the Taylor expansion of $f(x)$ contributes a term that is of the first order in t. The reason is that the *variance* of the increments of x is *linear* in t. This is the feature that makes the calculus of Brownian motion so different from the usual calculus of non-random variables.

A similar calculation will show that

$$Var[y_t - y_0] = f'(x_0)^2 \sigma^2 t + \ldots$$

Let x denote a general starting position and $y = f(x)$. Consider the infinitesimal increment dy over the next infinitesimal time interval dt. We can use the above expressions replacing t bt dt and ignoring higher order terms in dt. Therefore dy has mean

$$E[dy] = [f'(x)\mu + \tfrac{1}{2}f''(x)\sigma^2]dt,$$

and variance

$$Var[dy] = f'(x)^2 \sigma^2 dt.$$

So y follows the general diffusion process defined by

$$dy = [f'(x)\mu + \tfrac{1}{2}f''(x)\sigma^2]dt + f'(x)\sigma dw. \tag{1.6}$$

This is Itô's Lemma in the form that will prove most useful for us.

A slight generalization is easily available: if $y = f(x, t)$, the Taylor expansion has an addition term in f_t, and

$$dy = [f_x(x,t)\mu + \tfrac{1}{2}f_{xx}(x,t)\sigma^2 + f_t(x,t)]dt + f_x(x,t)\sigma dw. \tag{1.7}$$

For a simple intuition, return to the discrete random walk formulation, and suppose x has zero trend, so $p = q = 1/2$. Now $E[\Delta x] = 0$, and as time passes the distribution of x merely spreads out with linearly increasing variance around an unchanging mean. From the standard intuition of risk-aversion, or from Jensen's inequality, we know that the sign of $E[\Delta y]$ depends on the curvature of the function f. A risk-averter will dislike the increase in risk, and a risk-lover will like it. Therefore $E[\Delta y]$ will be negative if f is concave (f'' is negative) and positive if f is convex (f'' is positive). Figure 1.2 shows the latter case. We have

$$E[\Delta y] = \tfrac{1}{2} f(x + \Delta h) + \tfrac{1}{2} f(x - \Delta h) - f(x)$$
$$= \tfrac{1}{2} [f'(x)\Delta h + \tfrac{1}{2} f''(x)(\Delta h)^2 + \ldots]$$
$$+ \tfrac{1}{2} [-f'(x)\Delta h + \tfrac{1}{2} f''(x)(\Delta h)^2 + \ldots]$$
$$= \tfrac{1}{2} f''(x)(\Delta h)^2 + \ldots = \tfrac{1}{2} f''(x)\sigma^2 \Delta t + \ldots$$

Regarding Δt as an infinitesimal dt, this is exactly the same as the additional term in (1.6) that ordinary calculus would not have led us to expect. Thus Itô's Lemma is basically a consequence of Jensen's Inequality when we take into account the particular relation between the space steps and the time intervals of Brownian motion. Readers can now do the slightly messier case where $\mu \neq 0$ and get an exact correspondence with (1.6).

For more rigorous treatments, based on Itô integrals, see Harrison (1985, chapter 4), Karlin and Taylor (1981, chapter 15).

1.3. Geometric Brownian motion

Now suppose x follows the Brownian motion (1.1), and let $X = e^x$. Itô's Lemma gives

$$E[dX] = [e^x \mu + \tfrac{1}{2} e^x \sigma^2] dt = X[\mu + \tfrac{1}{2} \sigma^2] dt,$$

and

$$Var[dX] = [e^x]^2 \sigma^2 dt = X^2 \sigma^2 dt.$$

Therefore the process of X can be written

$$dX/X = [\mu + \tfrac{1}{2} \sigma^2] dt + \sigma dw.$$

This is called a geometric or proportional Brownian motion. It is particularly useful in economics because it provides a good first approx-

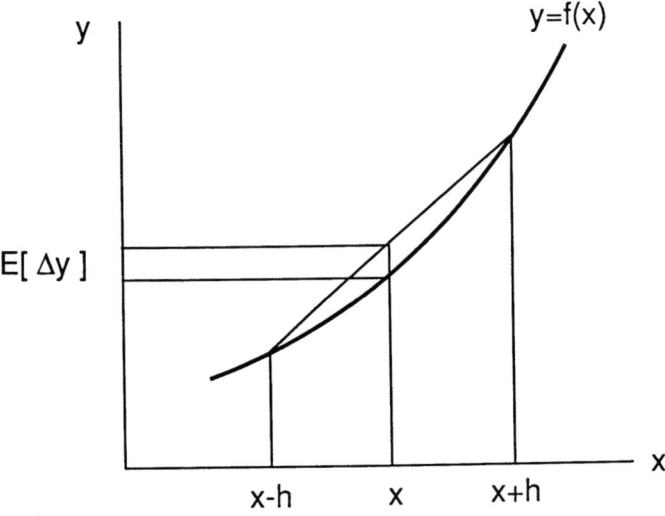

FIGURE 1.2 Itô's Lemma.

imation of the dynamics of exchange rates, prices of natural resources, and more generally many asset prices.

Conversely, if X follows the geometric Brownian motion

$$dX/X = \nu dt + \sigma dz, \tag{1.8}$$

then using Itô's Lemma we find that $x = \ln X$ follows the ordinary or absolute Brownian motion

$$dx = [\nu - \tfrac{1}{2}\sigma^2]dt + \sigma dw.$$

We will often have occasion to use this correspondence between geometric and absolute Brownian motions in what follows. Note that when $x = \ln X$, the trend coefficients on the right hand sides of the expressions (1.1) for dx and (1.8) for dX/X differ by $\tfrac{1}{2}\sigma^2$. Thus $d\ln X \neq dX/X$; this is the Jensen-Itô effect discussed above. The logarithm is a concave function, and therefore $d\ln X < dX/X$, and a calculation shows that $\tfrac{1}{2}\sigma^2 dt$ is just the right difference.

Suppose X follows the geometric Brownian motion (1.8), starting at $t = 0$ in the known position X_0. Let $x = \ln X$ and $x_0 = \ln X_0$. We know that x follows the absolute Brownian motion (1.1) with $\mu = \nu - \tfrac{1}{2}\sigma^2$. Then at any positive time t, $x_t = \ln X_t$ is normally distributed with mean $(x_0 + \mu t)$ and variance $\sigma^2 t$. In other words, X_t has a lognormal distribution with parameters $(x_0 + \mu t)$ and $\sigma\sqrt{t}$. For such a distribution, it can be shown that

$$E[X_t] = \exp(x_0 + \mu t + \tfrac{1}{2}\sigma^2 t) = X_0 e^{\nu t}.$$

This is a special case of the formula for a general exponential; see equation (2.2) in Section 2.1 below, see also Aitchison and Brown (1957, eq. 2.6). We see once again the Jensen–Itô effect at work; the exponential is a convex function, therefore

$$E[X_t] = E[\exp x] > \exp E[x_t] = \exp(x_0 + \mu t).$$

These repetitious details may seem boring, but they help fix the basics of the Itô calculus in the beginner's mind and thereby reduce the risk of many subsequent errors. Readers can get further practice by finding the stochastic process for $(1/X)$ when X follows the geometric Brownian motion (1.8). This is useful, for example, when converting formulas based on the dollar-yen exchange rate to ones using the yen-dollar rate.

1.4. Some generalizations

An obvious generalization of the Brownian motion (1.1) is obtained by letting the trend coefficient μ and the volatility coefficient σ depend on the current state x and also on time; thus

$$dx = \mu(x,t)dt + \sigma(x,t)dw. \tag{1.9}$$

A process whose trend and volatility coefficients are functions of the current state is often called a diffusion process; when they are functions of time as well, it is sometimes called an Itô process.

The geometric Brownian motion of Section 1.3 is an important special case. We can cast (1.8) in the diffusion process from (1.9) by writing

$$dX = \mu X dt + \sigma X dw,$$

so the coefficients are both proportional to the current state.

In some economic applications we need processes that revert toward some central level $\bar{x}$ of the state variable. For example, there may be an equilibrating force on prices. Now the trend coefficient has a sign opposite to that of $(x - \bar{x})$. When the mean reversion is linear, we have

$$dx = -\theta(x - \bar{x})dt + \sigma dw, \qquad (1.10)$$

where θ is some positive constant.

Finally, several variables x_i for $i = 1, 2, \ldots m$ may follow Brownian motion, their volatility components being linear combinations of independent standard Wiener processses w_j for $j = 1, 2, \ldots n$:

$$dx_i = \mu_i dt + \sum_{j=1}^{n} a_{ij} dw_j.$$

Then

$$E[dx_i] = \mu_i dt,$$

and

$$Var[dx_i] = \sum_{j=1}^{n} (a_{ij})^2 dt, \qquad Cov[dx_i, dx_k] = \sum_{j=1}^{n} a_{ij} a_{kj} dt.$$

This allows quite general correlation between the different dx_i. Such multi-variable processes are important in financial economics, but I will not develop these more advanced applications here.

2. DISCOUNTED PRESENT VALUES

Here we consider an economic unit, such as a firm, in a dynamic stochastic setting. Its state at time t is given by a state variable x_t that follows a Brownian motion with exogenous parameters μ and σ. There is a net flow payoff $f(x_t)$, such as profit or dividend, that depends on the state x_t. The expected present value $F(x)$ of the payoff starting at a given initial position $x_0 = x$, and using an exogenously specified discount rate ρ, is defined by

$$F(x) = E\left\{\int_0^{\infty} f(x_t)e^{-\rho t} dt \,\middle|\, x_0 = x\right\} \qquad (2.1)$$

Ultimately we will be interested in controlling or regulating the motion of x_t to optimize such expected present values net of the cost of control. To this end, we begin by evaluating $F(x)$ explicitly when

$f(x)$ has some particularly simple functional forms such as exponentials and polynomials. Then we can get power series expressions for $F(x)$ when $f(x)$ is any analytic function.

2.1. Present values for exponential and polynomials

Here we consider the special case when the flow payoff has the form

$$f(x) = \exp(\lambda x).$$

The discounted present value $F(x)$ will be finite when λ is in a certain range, which will be found in course of the calculation.

Starting from the initial value $x_0 = x$, the random position x_t at time t has a normal distribution with mean $(x + \mu t)$ and variance $\sigma^2 t$. Then

$$E[\exp(\lambda x_t) | x_0 = x] = \exp[\lambda(x + \mu t) + \tfrac{1}{2}\lambda^2 \sigma^2 t].$$

Note the Jensen–Itô effect at work for the convex exponential function yet again.

(Some readers may know this as a standard formula for the lognormal distribution, e.g. Aitchison and Brown (1957, eq. 2.6). Others may recognize it as the moment generating function for a normal distribution, e.g. Weatherburn (1946, p. 54). The rest can either accept the result as stated, or follow the explicit derivation below. Suppose a random variable x is normal with mean m and standard deviation s. Then

$$\begin{aligned}
E[\exp(\lambda x)] &= \frac{1}{s\sqrt{2\pi}} \int_{-\infty}^{\infty} \exp(\lambda x) \exp\left[-\frac{(x-m)^2}{2s^2}\right] dx \\
&= \frac{\exp(\lambda m)}{s\sqrt{2\pi}} \int_{-\infty}^{\infty} \exp\left[\lambda y - \frac{y^2}{2s^2}\right] dy \\
&= \frac{\exp[\lambda m + \tfrac{1}{2}\lambda^2 s^2]}{s\sqrt{2\pi}} \int_{-\infty}^{\infty} \exp\left[-\frac{(y - \lambda s^2)^2}{2s^2}\right] dy \\
&= \exp[\lambda m + \tfrac{1}{2}\lambda^2 s^2]. \quad (2.2)
\end{aligned}$$

The first line is just the definition of the expectation using the standard formula for the normal density. The second line transforms this to a new variable $y = x - m$ which is also normally distributed. The third line completes a square, while the fourth uses the fact that $y - \lambda s^2$ is also normal so its density integrates out to unity. The expectation of $\exp(\lambda x_t)$ is then just a special case of (2.2) with $m = x + \mu t$ and $s^2 = \sigma^2 t$.)

Using this expectation, we have the present value

$$F(x) = \int_0^\infty E[\exp(\lambda x_t)|x_0 = x] \exp(-\rho t) dt$$

$$= \exp(\lambda x) \int_0^\infty \exp[-(\rho - \lambda\mu - \tfrac{1}{2}\lambda^2\sigma^2)t] dt$$

$$= \exp(\lambda x)/(\rho - \lambda\mu - \tfrac{1}{2}\lambda^2\sigma^2). \qquad (2.3)$$

The integral converges provided the denominator is positive.

Such a convergence condition will appear repeatedly in such calculations. Therefore it is useful to give it a uniform notation and interpretation at the outset. For this purpose, define the function

$$\phi(\xi) \equiv \rho - \xi\mu - \tfrac{1}{2}\xi^2\sigma^2, \qquad (2.4)$$

which I shall label the Fundamental Quadratic of Brownian motion. Let us examine its properties. Since the coefficient of the quadratic term is negative, $\phi(\xi)$ goes to $-\infty$ as ξ goes to $\pm\infty$. We assume that the discount rate ρ is positive, which is the natural economic condition to ensure a finite present value for a constant flow. With $\phi(0) = \rho > 0$, the quadratic equation $\phi(\xi) = 0$ has two real roots, one negative, say $-\alpha$, and the other positive, say β. Then $\phi(\xi)$ is positive when ξ is in

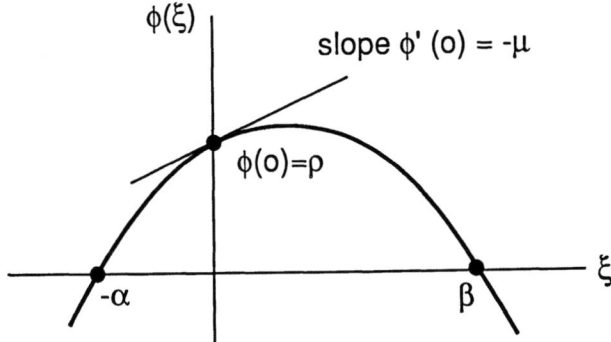

FIGURE 2.1 The Fundamental Quadratic.

the interval $(-\alpha, \beta)$, and negative outside it. Figure 2.1 shows this graphically.

Thus the condition for the convergence of the integral in (2.3) amounts to requiring that λ lies in the interval $(-\alpha, \beta)$. This includes $\lambda = 0$, a fact which will prove useful soon.

We can use the formula (2.3) above to derive an expression for $F(x)$ when $f(x)$ is any integer power x^n. Begin by observing that for the exponential flow function,

$$F(x) = E\left[\int_0^\infty \sum_{n=0}^\infty \frac{1}{n!}(\lambda x_t)^n e^{-\rho t} dt \,|\, x_0 = x\right]$$

$$= \sum_{n=0}^\infty \frac{\lambda^n}{n!} E\left[\int_0^\infty x_t^n e^{-\rho t} dt \,|\, x_0 = x\right]$$

Now we can expand the right hand side of (2.3) in powers of λ. Since the resulting power series must equal the series just above for all λ in a neighborhood that includes $\lambda = 0$, we can equate the coefficients of λ^n on the two sides. This yields an expression for the expected present value of the nth power of x.

To expand the right hand side of (2.3), note that

$$\exp(\lambda x) = \sum_{n=0}^\infty \frac{1}{n!} \lambda^n x^n,$$

and

$$[\rho - \mu\lambda - \tfrac{1}{2}\sigma^2\lambda^2]^{-1} = \frac{1}{\rho} \sum_{n=0}^\infty \rho^{-n}(\mu\lambda + \tfrac{1}{2}\sigma^2\lambda^2)^n.$$

For each n, the nth power of the sum of terms in λ and λ^2 in the second of these series must be written out using the binomial theorem. Then the two series can be multiplied together and the coefficients of like powers of λ collected to express the product as a single power series. This is tedious to do for higher powers, and in Section 2.5 we shall develop an alternative approach that is simpler. But the first few powers are not too hard, and yield the following results for the discounted present values of x and x^2:

$$E\int_0^\infty x_t e^{-\rho t} dt = \frac{\mu}{\rho^2} + \frac{x}{\rho}, \tag{2.5}$$

$$E\int_0^\infty x_t^2 e^{-\rho t}dt = \left(\frac{\sigma^2}{\rho^2} + \frac{2\mu^2}{\rho^3}\right) + \frac{2\mu x}{\rho^2} + \frac{x^2}{\rho}. \tag{2.6}$$

This last expression will prove useful in the menu cost application of Section 4.5 below.

Finally, consider any analytic $f(x)$ with the power series representation

$$f(x) = \sum_{n=0}^\infty f_n x^n,$$

assumed uniformly convergent for all x. Having found the expected present values of all integer powers, we can integrate term by term to find the $F(x)$ corresponding to this analytic $f(x)$. Thus we can in principle complete the calculation of present values for most common functions of Brownian motion.

2.2. Present values for powers of geometric Brownian motion

Next suppose X follows the geometric Brownian motion (1.8), namely

$$dX/X = \nu dt + \sigma dw.$$

We want to find the expected present value when the flow payoff is $g(X) = X^\lambda$. Note that $x = \ln(X)$ follows the Brownian motion

$$dx = (\nu - \tfrac{1}{2}\sigma^2)dt + \sigma dw,$$

and $X^\lambda = \exp(\lambda x)$. Using (2.3) for this, we have

$$E\left\{\int_0^\infty X_t^\lambda e^{-\rho t}dt\right\} = \exp(\lambda x)/[\rho - (\nu - \tfrac{1}{2}\sigma^2)\lambda - \tfrac{1}{2}\sigma^2\lambda^2]$$

$$= X^\lambda/[\rho - \nu\lambda - \tfrac{1}{2}\sigma^2\lambda(\lambda - 1)]. \tag{2.7}$$

provided the integral converges, for which we need the denominator to be positive.

The convergence condition is now defined in terms of a slightly different quadratic

$$\psi(\xi) \equiv \rho - \nu\xi - \tfrac{1}{2}\sigma^2\xi(\xi - 1). \tag{2.8}$$

I make two economically natural assumptions that help locate the roots: $\rho > 0$, which ensures convergence of the expected present value of a constant flow, and $\rho > \nu$, which guarantees the convergence of the expected present value of X_t itself. Then the roots of the quadratic (2.8) are $-\gamma < 0$ and $\delta > 1$. For convergence of (2.7) we require that λ should lie in the interval $(-\gamma, \delta)$.

Note that the expected present value is finite for all integer powers of absolute Brownian motion, but not for all those of geometric Brownian motion. The reason is that a power of geometric Brownian motion is like an exponential of absolute Brownian motion. Applying Itô's Lemma again, we find that $Y = X^\lambda$ follows the geometric Brownian motion

$$dY/Y = [\lambda \nu + \tfrac{1}{2}\lambda(\lambda - 1)\sigma^2]dt + \lambda \sigma dw.$$

(The reader should derive this to continue the process of acquiring familiarity with Itô's Lemma.) For convergence of the expected present value of Y, the discount rate must exceed the trend growth rate of Y. Inspection shows this to be just the condition $\psi(\lambda) > 0$.

Note also the relationship between the two fundamental quadratics, (2.4) for an absolute Brownian motion $\{x_t\}$ with parameters (μ, σ) and (2.8) for a geometric Brownian motion $\{X_t\}$ with parameters (ν, σ). If in fact $x = \ln(X)$, then the two trend parameters will be related to each other by $\mu = \nu - 1/2\sigma^2$. Substituting in (2.8), we will get

$$\psi(\xi) = \rho - (\mu + \tfrac{1}{2}\sigma^2)\xi - \tfrac{1}{2}\sigma^2\xi(\xi - 1)$$
$$= \rho - \mu\xi - \tfrac{1}{2}\sigma^2\xi^2 = \phi(\xi),$$

and then the roots will also correspond, with $\alpha = \gamma$ and $\beta = \delta$.

2.3. A basic differential equation for present value

Let us return to the absolute Brownian motion x and the flow payoff $f(x)$, and consider an alternative characterization of the expected present value $F(x)$ defined in (2.1). For this purpose, we split the integral into the contribution over the initial infinitesimal time interval from 0 to dt, and the integral from dt to ∞. In spirit this is exactly like the decomposition of an intertemporal objective in dymanic programming; the only difference is that we have no control variables here. Now at time dt the state variable will attain a value $(x + dx)$ that is not known at time 0, although its distribution is. The integral from there on will be just $F(x + dx)$, which must be discounted back to time 0. Thus we have

THE ART OF SMOOTH PASTING 15

$$F(x) = f(x)dt + e^{-\rho dt}E[F(x + dx)].$$

Note that this is already an approximation in regarding $f(x)$ as constant over the small interval dt. The resulting error in $f(x)dt$ is of order dt^2, and therefore negligible in the limit as dt goes to zero. We further simplify the expression, continuing to ignore terms that are small relative to dt.

$$F(x) = f(x)dt + (1 - \rho dt)(F(x) + \{E[F(x + dx)] - F(x)\})$$
$$= f(x)dt + F(x) - \rho F(x)dt + \{E[F(x + dx)] - F(x)\}.$$

Therefore

$$\rho F(x)dt = f(x)dt + \{E[F(x + dx)] - F(x)\}$$
$$= f(x)dt + E[dF]. \quad (2.9)$$

This is an arbitrage equation. Think of the entitlement to the flow payoffs as a capital asset; $F(x)$ is its value. Contemplate holding this asset over the period $(t, t + dt)$. This yields a dividend $f(x)dt$, and an expected capital gain $E[dF]$. The sum of these two should equal the normal return $\rho F(x)dt$.

· By Itô's Lemma,

$$E[dF] = \mu F'(x)dt + \tfrac{1}{2}\sigma^2 F''(x)dt.$$

Substituting into (2.9) and dividing by dt, we get

$$\tfrac{1}{2}\sigma^2 F''(x) + \mu F'(x) - \rho F(x) + f(x) = 0. \quad (2.10)$$

2.4. Derivation by discrete approximation

We regarded Brownian motion as the limit of a discrete random walk, and we can also derive the differential equation (2.10) by that approach. This may be easier for some readers to follow, and similar methods will prove useful in later sections where we consider controlling or regulating the motion. Therefore I present this derivation before turning to the solution of the equation.

Label the discrete points in the x space by i, and the discrete time periods by j. Let i_j denote the position of the particle at time j; future positions are of course random variables given our initial information at $j = 0$. Then the expected present value can be written as

$$F(i) = E\left\{\sum_{j=0}^{\infty} f(i_j)\Delta t e^{-j\rho\Delta t} \,\bigg|\, i_0 = i\right\}$$

After the first step, the same problem restarts with a new initial state i_1, which from the time 0 perspective can be either $(i+1)$ with probability p or $(i-1)$ with probability q. Thus the expectation on the right hand side becomes

$$F(i) = f(i)\Delta t + e^{-\rho\Delta t}[pF(i+1) + qF(i-1)].$$

Now expand the right hand side, ignoring terms of higher order than Δt. Note that

$$e^{-\rho\Delta t} = 1 - \rho\Delta t + \ldots$$

Next, using definition (1.5') of p and q, and the relation (1.4) between the stepsize Δh and the time interval Δt, we get

$$\begin{aligned}pF(i+1) + qF(i-1) &= \tfrac{1}{2}[1 + (\mu/\sigma)\sqrt{\Delta t}]F(x+\Delta h) \\ &\quad + \tfrac{1}{2}[1 - (\mu/\sigma)\sqrt{\Delta t}]F(x-\Delta h) \\ &= \tfrac{1}{2}[1 + (\mu/\sigma)\sqrt{\Delta t}]\,[F(x) + F'(x)\Delta h \\ &\quad + \tfrac{1}{2}F''(x)(\Delta h)^2 + \ldots] \\ &\quad + \tfrac{1}{2}[1 - (\mu/\sigma)\sqrt{\Delta t}]\,[F(x) \\ &\quad - F'(x)\Delta h + \tfrac{1}{2}F''(x)(\Delta h)^2 + \ldots] \\ &= F(x) + \mu F'(x)\Delta t + \tfrac{1}{2}\sigma^2 F''(x)\Delta t + \ldots\end{aligned}$$

Substituting and simplifying yields the same equation as (2.10) above.

2.5. The general solution

The differential equation (2.10) is linear (in the dependent variable and its derivatives). Therefore its general solution is the sum of two parts: any solution of the equation as a whole (the particular integral) and the general solution of the homogeneous part of the equation with the term $f(x)$ omitted (the complementary function).

To find the complementary function, write the homogeneous part of the equation:

$$\tfrac{1}{2}\sigma^2 F''(x) + \mu F'(x) - \rho F(x) = 0.$$

Its general solution can be expressed as a linear combination of two independent solutions. If we try solutions of the form $\exp(\xi x)$, we get

THE ART OF SMOOTH PASTING 17

$$\exp(\xi x)\left[\tfrac{1}{2}\sigma^2\xi^2 + \mu\xi - \rho\right] = 0.$$

Since the exponential is always positive, this holds if and only if

$$\rho - \mu\xi - \tfrac{1}{2}\sigma^2\xi^2 = 0.$$

This is just the fundamental quadratic $\phi(\xi)$ introduced in (2.4) above. So ξ must equal $-\alpha$ or β, the two roots. The two roots are distinct, since $\rho > 0$ ensures that the roots have opposite signs. Therefore the two solutions $e^{-\alpha x}$ and $e^{\beta x}$ are independent, and the general solution is

$$Ae^{-\alpha x} + Be^{\beta x}, \qquad (2.11)$$

where A and B are undetermined constants.

Finding a particular solution to the full equation (2.10) is often an art, but for the exponential and polynomial forms we tried before, there are obvious choices.

Begin with the exponential case. When $f(x) = \exp(\lambda x)$, the form $F(x) = K\exp(\lambda x)$ suggests itself. Substituting in (2.10), we have

$$K(\tfrac{1}{2}\sigma^2\lambda^2 + \mu\lambda - \rho)\exp(\lambda x) + \exp(\lambda x) = 0,$$

or

$$K = -1/(\tfrac{1}{2}\sigma^2\lambda^2 + \mu\lambda - \rho) = 1/\phi(\lambda),$$

using the notation of our fundamental quadratic (2.4). When λ lies between the two roots $-\alpha$ and β of the quadratic, $\phi(\lambda)$ is positive. Then K is also positive.

Combining this particular solution and the earlier complementary function (2.11), the general solution for the expected present value in the exponential case becomes

$$F(x) = \frac{1}{\phi(\lambda)}e^{\lambda x} + Ae^{-\alpha x} + Be^{\beta x}, \qquad (2.12)$$

where the constants A and B remain to be determined.

In fact a simple argument based on the definition (2.1) shows that when the flow $f(x)$ is the exponential $\exp(\lambda x)$, the expected present values $F(x)$ must be a multiple of $\exp(\lambda x)$. To prove this, consider $F(x + h)$ for any h. This means the initial point of the process x_t is now taken as $x_0 = x + h$. For each path of the Brownian particle starting at x, there is an equiprobable parallel path starting at $x + h$.

Along the latter path, the flow is always $\exp(\lambda h)$ times that along the former. Then the same must hold for the expected present values. In particular,

$$F(h) = e^{\lambda h} F(0).$$

To give a somewhat more formal argument, define $y_t = x_t - h$, and consider the stochastic process y_t. This is also a Brownian motion with

$$dy = dx = \mu\, dt + \sigma\, dw,$$

and the initial position $y_0 = (x + h) - h = x$. The flow benefit can be written

$$f(x_t) = e^{\lambda x_t} = e^{\lambda h} e^{\lambda y_t} = e^{\lambda h} f(y_t).$$

Integrating over time and taking expectations, we get

$$F(x + h) = e^{\lambda h} E\left\{\int_0^\infty f(y_t) e^{-\rho t}\, dt \,\bigg|\, y_0 = x\right\}$$
$$= e^{\lambda h} F(x).$$

Subtracting $F(x)$ from both sides, dividing by h, and letting h go to zero, we get

$$\lim_{h \to 0} \frac{F(x + h) - F(x)}{h} = \lim_{h \to 0} \frac{\exp(\lambda h) - 1}{h} F(x),$$

or

$$F'(x) = \lambda F(x).$$

Then $F(x)$ must take the form $K\exp(\lambda x)$ where K is a constant.

The general solution (2.12) above was a combination of three terms, of which only the first, corresponding to the particular solution we guessed initially, had the right exponential form. Thus that guess is in fact the full solution, and both constants A and B in the complementary function (2.11) are zero.

In the same way we will be able to guess the right solution as our particular integral for many simple forms of the flow function. The complementary function will play a more important role in the next section, where it will capture the modification to expected present values that is caused by the barriers.

Next consider the polynomial case. When $f(x) = x^n$ for a positive integer n, a natural guess for the particular integral is

$$F(x) = \sum_{m=0}^{n} a_m x^m. \tag{2.13}$$

Substituting this in (2.10), we get

$$\tfrac{1}{2}\sigma^2 \sum_{m=2}^{n} m(m-1)a_m x^{m-2} + \mu \sum_{m=1}^{n} m a_m x^{m-1}$$

$$- \rho \sum_{m=0}^{n} a_m x^m + x^n = 0.$$

Collecting like powers of x together, and equating the coefficient of each separately to zero since the equation must hold as an identity in x, we find

$$a_n = 1/\rho, \qquad a_{n-1} = n\mu/\rho^2,$$

and for $m = 0, 1, 2, \ldots (n-2)$, the recursive relation

$$\rho a_m = (m+1)\mu a_{m+1} + \tfrac{1}{2}(m+1)(m+2)\sigma^2 a_{m+2}. \tag{2.14}$$

This determines all the coefficients a_m. Once again we can verify that the expected present value cannot have any contribution from the exponentials of the complementary function (2.11), and therefore (2.13) is the full solution. This method, while needing a recursive solution, is simpler than the power series expansion we developed in Section 2.1 above. Readers can explicity write out the results using the above recursion for $n = 1$ and $n = 2$, and check their consistency with the formulas (2.5) and (2.6) derived above by the power series method.

2.6. Differential equation for geometric Brownian motion

Now suppose the underlying variable is X, and it follows the proportional or geometric Brownian motion (1.8). Given a flow cost function $g(X)$, we want to find

$$G(X) = E\left\{\int_0^\infty g(X_t)e^{-\rho t}dt \,\bigg|\, X_0 = X\right\} \tag{2.15}$$

Proceeding exactly as before, we get the arbitrage equation

$$\rho G(X)dt = g(X)dt + E[dG],$$

and Itô's Lemma gives

$$E[dG] = \nu XG'(X)dt + \tfrac{1}{2}(\sigma X)^2 G''(X)dt.$$

Therefore the basic differential equation for the case of geometric Brownian motion is

$$\tfrac{1}{2}\sigma^2 X^2 G''(X) + \nu X G'(X) - \rho G(X) + g(X) = 0. \quad (2.16)$$

The complementary function (the general solution of the homogeneous part) of (2.16) is easily seen to be

$$CX^{-\gamma} + DX^{\delta}, \quad (2.17)$$

where $-\gamma$ and δ are the roots of the fundamental quadratic (2.8) for geometric Brownian, and C, D are constants to be determined.

Once again the particular integral must be guessed. In Section 2.2 above we considered a flow function of the form $g(X) = X^\lambda$. Using the differential equation method, the natural guess for the corresponding expected present value $G(X)$ is KX^λ where K is a constant to be determined. Substituting in (2.16) we find

$$K[\tfrac{1}{2}\sigma^2 \lambda(\lambda - 1) + \nu\lambda - \rho]X^\lambda + X^\lambda = 0,$$

or

$$K = -1/[\tfrac{1}{2}\sigma^2 \lambda(\lambda - 1) + \nu\lambda - \rho] = 1/\psi(\lambda),$$

where $\psi(\lambda)$ is the fundamental quadratic for the geometric case, and is positive when λ lies between the two roots $-\lambda$ and δ.

An argument similar to that we made above in Section 2.5 for the case of absolute Brownian motion and an exponential flow shows in the present case that $G(X)$ must take the form KX^λ, in other words it inherits the homogeneity of degree λ from $g(X)$. Then the full solution is just the particular integral we guessed; the constants C and D in the complementary function (2.17) are both zero.

We can change variables to transform geometric Brownian motion into an absolute one, and this gives an alternative way to find expected present values. Define $x = \ln(X)$, which then follows the absolute Brownian motion (1.1) with $\mu = \nu - \tfrac{1}{2}\sigma^2$. Now let $f(x) = g(e^x)$, use the earlier (2.1) to get $F(x)$, and then set $G(X) = F(\ln(X))$. Let us check that the transformation yields the same ultimate result as the direct analysis of geometric Brownian motion above. Note that $F(x) = G(e^x)$, so

$$F'(x) = e^x G'(e^x) = XG'(X)$$

and

$$F''(x) = (e^x)^2 G''(e^x) + e^x G'(e^x) = X^2 G''(X) + XG'(X).$$

Substituting in (2.10), we find

$$\begin{aligned} 0 &= \tfrac{1}{2}\sigma^2[X^2 G''(X) + XG'(X)] + \mu XG'(X) - \rho G(X) + g(X) \\ &= \tfrac{1}{2}\sigma^2 X^2 G''(X) + [\mu + \tfrac{1}{2}\sigma^2]XG'(X) - \rho G(X) + g(X) \\ &= \tfrac{1}{2}\sigma^2 X^2 G''(X) + \nu XG'(X) - \rho G(X) + g(X), \end{aligned}$$

which is just (2.16). Thus the two approaches are mutually consistent. And the transformation of the differential operators casts a new light on the nonlinearity leading to Itô's Lemma from a somewhat different angle.

2.7. General diffusion processes

If x follows the general diffusion process (1.9) rather than the simple Brownian motion (1.1), the expected value function $F(x)$ and the flow function $f(x)$ are linked by a differential equation that is a natural generalization of (2.10), namely

$$\tfrac{1}{2}\sigma(x)^2 F''(x) + \mu(x)F'(x) - \rho F(x) + f(x) = 0. \quad (2.18)$$

Unfortunately, the solution of this is not the corresponding simple generalization of the solution of (2.10). The complementary function is specific to each case depending on the functional forms of $\mu(x)$ and $\sigma(x)$, and does not have any simple form like (2.11). Analytical solution is possible only in very special cases. The geometric case was discussed above. for the linearly mean-reverting motion (1.10), a power series solution related to the Confluent Hypergeometric Function is available; an example is developed in Section 5.1 below. But many other applications require numerical solution of the differential equation.

If x follows the Itô process (1.9) whose parameters depend on time as well as the state x, or if the flow payoff is a function $f(x, t)$ likewise, or the process ends at a given time T so the time remaining to the end of the horizon matters, then we must allow the expected present value to depend on time too. Applying Itô's Lemma to $F(x, t)$ introduces an additional term for the time derivative as in (1.7), and then the basic equation becomes a *partial* differential equation

$$\tfrac{1}{2}\sigma(x,t)^2 F_{xx}(x,t) + \mu(x,t)F_x(x,t) - \rho F(x,t) + F_t(x,t) + f(x,t) = 0. \tag{2.19}$$

The solution of this is much harder, and typically needs numerical methods. I shall mention only one application of this very briefly in Section 5.2.

3. BARRIERS

The above calculations of expected present values assumed that the Brownian motion particle x was free to range over the entire real line $(-\infty, \infty)$; similarly in the case of geometric Brownian motion, X could range over $(0, \infty)$. In practice there are restrictions on the range. Sometimes these arise naturally. For example the output price facing one firm in a competitive industry is bounded above, because new firms will enter if the price rises beyond a certain point. At other times the restrictions are exogenously imposed, as in the case of a government-imposed agricultural price floor. Finally, and most important for our purpose here, the restrictions arise endogenously through purposive optimal control of the stochastic process.

In this section I shall start with some specified restrictions on the process, and show how their effects on expected present values can be computed. That will pave the way for the next section, where I characterize the optimal choice of these restrictions when the aim is to maximize the expected present value of a flow benefit net of the cost of exercising control over the Brownian motion.

The restrictions are called *barriers*. They can constrain upward or downward movements of x, and are of two types, *absorbing* and *reflecting*.

An upper absorbing barrier at b means that the process (1.1) is allowed to proceed unhindered so long as $x_t < b$, but if ever $x_t = b$, the process is terminated. That might be the end of our planning horizon, or merely the end of the movement of x, so that it remains at b for ever after. Sometimes such an absorbed process might be immediately restarted at a point $c < b$. Lower and two sided absorbing barriers are defined in obvious ways.

An upper reflecting barrier at b means that the process (1.1) is allowed to proceed unhindered so long as $x_t = b$, but if ever $x_t = b$, and the next increment dx is positive, then the sign of this increment is reversed, as if the particle were reflected in a mirror placed at b. Once again, lower and two-sided barriers are defined analogously. We can even have an absorbing barrier on one side and a reflecting barrier on the other side.

Now consider a process with barriers for x, take a flow payoff function $f(x)$, and define the expected present value function $F(x)$ as in (2.1). When there were no barriers, we took two approaches to finding $F(x)$. The first approach was direct. The distribution of x_t given x_0 was known and simple, namely normal, so expected values of functions of x_t could be found relatively easily. When there are barriers, between the starting time 0 and the instant in question t the particle might have been reflected at barriers any number of times, or been absorbed with positive probability. The distribution of x_t conditional on x_0 is much more complicated. In Section 6.3 I shall begin to develop the methods that yield this distribution. But for our present purpose, the second and indirect approach in Section 2 proves simpler. We show that $F(x)$ satisfies the same basic differential equation in the interior of the region of variation of x, and certain end-point conditions at the barriers. Finding $F(x)$ then amounts to solving the differential equation subject to the end-point conditions.

For a more rigorous treatment of expected present values with barriers, see Harrison (1985, pp. 44-49).

3.1. Basic differential equation

Suppose the process moves between barriers located at a and b. For the moment it makes no difference whether the barriers are reflecting or absorbing, and $a = -\infty$ or $b = \infty$ are permissible. Choose the initial point $x_0 = x$ in the interior of the range $[-a, b]$. Over an infinitesimal time interval dt, the probability of x_t reaching either barrier is negligible. Therefore the arbitrage argument of Section 2.3 remains valid, and the basic differential equation (2.10) holds. The derivation of Section 2.4, based on discretization of the x space, is also valid so long as the length of each step Δh is chosen sufficiently small.

Even when the process has barriers, the flow function $f(x)$ is generally defined over the full unrestricted range $(-\infty, \infty)$ of the x space. Even if it is not, we can extend its definition in some simple way over the full range. Let $F_0(x)$ denote the expected present value of $f(x_t)$ as defined in (2.1) above, and computed *ignoring the barriers*. The arbitrage argument of Section 2.3 (or the discrete version of Section 2.4) shows that $F_0(x)$ satisfies the same basic differential equation over the full range, and *a fortiori* over the range (a, b) between the barriers. Therefore we can choose $F_0(x)$ as a particular integral for (2.10). We already know it for some functional forms of special interest, and can in principle calculate it for any analytic function $f(x)$.

Our choice of $F_0(x)$, the expected present value ignoring barriers, as the particular integral gives a very nice economic interpretation to the solution. Since the particular integral is what the expected present value would have been in the absence of barriers, the remaining part, namely the complementary function, must equal the effect of the barriers. For example, a price ceiling cuts off the upside profit potential of a firm; its effect is simply captured by the appropriate term in the complementary function. In other instances where the barriers are chosen optimally, the complementary function is the addition to the expected present value that results from the ability to exercise such control.

Recalling the form (2.11) of the complementary function, write the general solution

$$F(x) = F_0(x) + Ae^{-\alpha x} + Be^{\beta x}. \tag{3.1}$$

The constants A and B must be determined using some other conditions on the problem. This is where the barriers come into play. If the process is absorbed or reflected at a or b, that implies certain conditions on $F(x)$ or its derivatives at the barriers. In the event $a = -\infty$ or $b = \infty$, there are corresponding conditions that arise from the asymptotic behavior of $F(x)$. The exact form of the conditions, and therefore of the solution, depends on the type and position of the barrier, and we examine various cases separately.

If x is free to range over the entire line $(-\infty, \infty)$, we already know that $F(x) = F_0(x)$. Then A and B must both be zero.

If x is not restricted on the lower side, but has an upper barrier at b, then we can get some information by considering what happens for very large negative values of x. Starting from such a value, the particle is unlikely to reach b in any reasonable future time. Then the unrestricted expected present value $F_0(x)$ should be a good approximation for $F(x)$. But with $\alpha > 0$, $e^{-\alpha x}$ goes to ∞ as x goes to $-\infty$. This would spoil the desired approximation unless $A = 0$. Thus we have determined one of the constants. The other, B, is fixed using endpoint conditions at the barrier b, which we will examine shortly.

Similarly, if x has a only lower barrier at a, then the inspection of $F(x)$ as x goes to ∞ gives $B = 0$, while A is fixed by conditions at the barrier. If there are barriers on both sides, then both A and B are fixed by the end-point conditions at the barriers.

3.2. Geometric Brownian motion

Now suppose the underlying variable is X, and it follows the proportional or geometric Brownian motion (1.8), with barriers at c and d. Let the flow function be $g(X)$. Extending its definition outside the barriers to the full range $(0, \infty)$ of X if necessary, let $G_0(X)$ be the expected present value of the flow *ignoring the barriers*. Then $G_0(X)$ satisfies the basic differential equation (2.16) for geometric Brownian motion, and serves as a particular integral over the restricted range (c, d). The general solution is

$$G(X) = G_0(X) + CX^{-\gamma} + DX^\delta, \qquad (3.2)$$

$-\gamma$ and δ are the roots of the fundamental quadratic (2.8), and C, D are constants to be determined by conditions that apply at the barriers, or from the asymptotic behavior of $G(X)$ in the absence of one or both barriers. I shall leave the details for the readers to fill in by analogy with those at the end of the previous section.

3.3. Stopping

Now let us return to the case of absolute Brownian motion, and begin the analysis of barriers, starting with an upper absorbing barrier. Suppose the barrier is placed at b. To set the stage for subsequent analysis of control, allow an exogenous terminal payoff $W(b)$. If x stays for ever at b after absorption, $W(b)$ may simply be the capitalized value of the constant flow payoff, $f(b)/\rho$. But other interpretations are also possible. If a cost k must be paid at the instant of absorption, we simply subtract it from $W(b)$ to make it $W(b) - k$.

The expression (2.1) for the expected present value must be modified to take account of the barrier and the terminal payoff. Thus

$$F(x) = E\left\{\int_0^{t(b)} f(x_t)e^{-\rho t}dt + e^{-\rho t(b)}W(b)\,\bigg|\,x_0 = x\right\} \quad (3.3)$$

where $t(b)$ is the first time the process reaches b starting at x, and of course it is a random variable given the initial information.

To find the conditions that hold at the barrier, we repeat the arbitrage calculation, but now starting at or near the barrier rather than at an interior point.

Revert to the discrete approximation to Brownian motion, with time intervals Δt and steps of length Δh. Starting at $(b - \Delta h)$, we have

$$F(b - \Delta h) = f(b - \Delta h)\Delta t + (1 - \rho\Delta t)[pW(b) + qF(b - 2\Delta h)].$$

Expanding the Taylor series, we get

$$F(b) - F'(b)\Delta h + \ldots = f(b)\Delta t + \ldots + \tfrac{1}{2}W(b)[1 - \rho\Delta t + \ldots]$$
$$+ \tfrac{1}{2}(1 - \rho\Delta t)[F(b) - 2F'(b)\Delta h + \ldots]$$

Collecting terms, dividing by Δh and taking limits, we find

$$F(b) = W(b). \quad (3.4)$$

This is sometimes called the Value Matching Condition.

3.4. Resetting

Here the process is allowed to follow (1.1) so long as $x < b$, but the instant x hits b, it is reset at $x = c < b$, and the process is restarted. The calculation is as above, except that $F(c)$ replaces $W(b)$, and we get a value matching condition $F(b) = F(c)$. If the resetting entailed a cost k, this would become $F(b) = F(c) - k$.

3.5. Reflection

Finally, suppose the process is reflected at an upper barrier b. Now starting at b we are sure to go to $(b - \Delta h)$, so

$$F(b) = f(b)\Delta t + (1 - \rho\Delta t)F(b - \Delta h)$$
$$= f(b)\Delta t + (1 - \rho\Delta t)[F(b) - F'(b)\Delta h + \ldots]$$

Cancelling $F(b)$ from both sides, the leading term on the right hand side becomes $-F'(b)\Delta h$; recall that it is of order $\sqrt{\Delta t} \gg \Delta t$. Dividing Δh and taking limits gives

$$F'(b) = 0. \qquad (3.5)$$

To repeat the point in a slightly different way, note that starting at b, the next step is sure to be downward through the distance $\Delta h = \sigma\sqrt{\Delta t}$. If $F'(b)$ were non-zero, this would make $dF = -\sigma F'(b)\sqrt{\Delta t}$. The capital gain term in the arbitrage equation (2.9) would be of order $\sqrt{\Delta t}$. But the normal return and dividend terms are of order Δt, and therefore relatively much smaller. Then the arbitrage equation would not hold. This contradiction proves that $F'(b) \neq 0$ is impossible.

Some people call any condition that pertains to the first-order derivatives of an expected present value of a function of Brownian motion a Smooth Pasting Condition. This would be their name for (3.5). Others prefer to reserve the term 'Smooth Pasting' for an optimality condition for termination or resetting; that is developed in the next chapter. Usually the context clarifies the usage. See also Malliaris and Brock (1982, p. 200).

If reflection is costly, with cost m per unit distance through which the particle is reflected, we substract $m\,\Delta h$ from the right hand side, and the condition becomes $F'(b) + m = 0$. At a lower reflecting barrier, a, similar analysis gives $F'(a) - m = 0$.

Similar conditions for geometric Brownian motion are easily derived, so there is no need to repeat the calculations.

3.6. Example: price ceiling

Consider a firm that produces a unit of output per unit time. The cost of production is W. The price P follows a geometric Brownian motion with parameters ν and σ. In this section I suppose that even when P falls below W so that the operating profit $(P - W)$ is negative, the firm must continue operation. This is sometimes required of public utilities or transport services. In Section 3.9 I consider an alternative where the firm can temporarily suspend operations.

Begin by supposing that there are no barriers on the price process. Then we can use the formula (2.7) for $\lambda = 0$ and 1 to get the expected present value of profits

$$G_0(P) = P/(\rho - \nu) - W/\rho. \tag{3.6}$$

Next suppose there is an upper reflecting barrier on the price process at $P = b$. This could be a ceiling imposed by the government, or the result of new entry of identical competitive firms. Then, following the method laid out in Section 3.2, the expression for the expected present value of profits is easily seen to be

$$G(P) = DP^\delta + P/(\rho - \nu) - W/\rho. \tag{3.7}$$

Note that the other term in the complementary function, $CP^{-\gamma}$, must be zero to ensure a finite expected value as P goes to zero. To determine the constant D, we use the condition (3.5), so

$$G'(b) = \delta D b^{\delta - 1} + 1/(\rho - \nu) = 0.$$

Then

$$D = -b^{1-\delta}/[\delta(\rho - \nu)] < 0.$$

Comparing the expressions (3.6) and (3.7), we see how the ceiling reduces the expected present value of profits by cutting off the upside potential. Substituting for D in (3.7), we get

$$G(P) = -\frac{P^\delta b^{1-\delta}}{\delta(\rho - \nu)} + \frac{P}{\rho - \nu} - \frac{W}{\rho}. \tag{3.8}$$

3.7. Example: exchange rate target zones

The following reduced form model is often used in exchange rate theory. Let s denote the logarithm of the exchange rate, and x the logarithm of the fundamental determinant of it, typically a variable such as money supply or the velocity of circulation of money. Then arbitrage-type considerations establish

$$s = x + \lambda E[ds]/dt, \tag{3.9}$$

where λ can be interpreted as the semi-elasticity of money demand. Let $\rho = 1/\lambda$. If an explosive bubble path is ruled out, we can integrate (3.9) to write

$$s_0 = \rho \int_0^\infty x_t e^{-\rho t} dt. \qquad (3.10)$$

This is formally very like the expected present value expression (2.1). Suppose x_t is a Brownian motion with parameters $\mu = 0$ and σ. Then, using (2.5), the right hand side of (3.10) becomes simply x_0. In the absence of any barriers of controls, the exchange rate just tracks the fundamentals.

Next suppose we wish to keep the exchange rate confined to the interval $(-z, z)$. We do this by confining the fundamental process to $(-b, b)$ by means of reflecting barriers at both ends. To see what this does to s, write $s = S(x)$ and substitute in (3.9). Using Itô's Lemma, we get

$$s = x + \tfrac{1}{2} \lambda \sigma^2 S''(x),$$

or

$$\tfrac{1}{2} \sigma^2 S''(x) - \rho S(x) + \rho x = 0.$$

This can be solved by familiar methods. The fundamental quadratic is

$$\phi(\xi) \equiv \rho - \tfrac{1}{2} \sigma^2 \xi^2 = 0,$$

with roots $\pm \beta = \pm \sqrt{2\rho}/\sigma$. The general solution is

$$S(x) = x + A e^{-\beta x} + B e^{\beta x}.$$

The two constants are determined using the 'smooth pasting' conditions at the two reflecting barriers:

$$S'(-b) = 0 = S'(b).$$

The result is

$$S(x) = x - \frac{1}{\beta} \frac{e^{\beta x} - e^{-\beta x}}{e^{\beta b} + e^{-\beta b}}. \qquad (3.11)$$

Finally, b must be chosen to attain the desired limits on s, so $S(b) = z$ defines b in terms of the given z. We do not need the explicit expression. The result is an S-shaped curve, with slope

$$S'(x) = 1 - \frac{e^{\beta x} + e^{-\beta x}}{e^{\beta b} + e^{-\beta b}}.$$

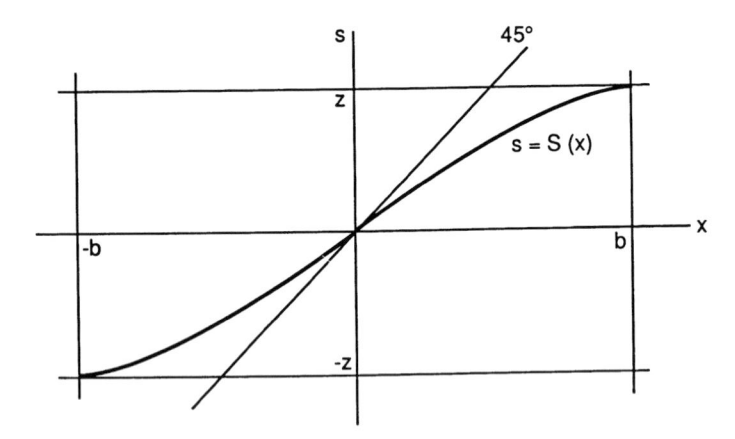

FIGURE 3.1 Exchange rate in target zone.

Note that the function $\exp(\beta x) + \exp(-\beta x)$ is symmetric and convex. Its minimum occurs at $x = 0$, the minimum value being 2. Therefore (i) $S'(0)$ lies between 0 and 1, and (ii) $S'(x)$ goes to 0 as x goes to $\pm b$. Then the function $S(x)$ giving the exchange rate in terms of the fundamental is S-shaped, and everywhere flatter than 1. In other words, shocks to the fundamental have a dampened effect on the exchange rate. This idea and the model originated with Krugman (1991); see Svensson (1992) for a lucid account of it and of the volume of subsequent research on target zones.

3.8. Transitional boundary

Sometimes the process encounters not a barrier, but a point of transition where either the parameters of the process, or the flow payoff function, or both, undergo a change. Let b be such a point of transition, and suppose

$$\mu = \mu_1, \sigma = \sigma_1, f(x) = f_1(x) \text{ when } x < b,$$
$$\mu = \mu_2, \sigma = \sigma_2, f(x) = f_2(x) \text{ when } x > b.$$

We can use separate differential equations like (2.10) on each side of b and obtain solutions $F_1(x)$ and $F_2(x)$. The former is valid for $-\infty < x < b$, and the latter for $b < x < \infty$. Asymptotic considera-

THE ART OF SMOOTH PASTING   31

tions allow us to get rid of the term in $\exp(-\alpha_1 x)$ from $F_1(x)$, and that in $\exp(\beta_2 x)$ from $F_2(x)$, where $-\alpha_i$, β_i denote the roots of the fundamental quadratic in the respective regions $i = 1, 2$. Note that $F_1(x)$ has no relevance to the right of b, so we cannot use the limiting consideration as x goes to ∞ for $F_1(x)$. Similarly $F_2(x)$ cannot be made to satisfy any asymptotic regularity requirement as $x \to -\infty$. Therefore each of $F_1(x)$ and $F_2(x)$ still contains one constant to be determined.

The condition that fixes the two remaining constants is that $F_1(x)$ and $F_2(x)$ should meet tangentially, or be 'smoothly pasted' together, at b. Thus

$$F_1(b) = F_2(b), \quad F_1'(b) = F_2'(b). \tag{3.12}$$

To see the intuition behind this, consider arbitrage starting at b. Use the discrete approximation. To the left of b, the process has steps of length $\Delta h_1 = \sigma_1 \sqrt{\Delta t}$, while to the right they are of length $\Delta h_2 = \sigma_2 \sqrt{\Delta t}$. Now if the values $F_1(b)$ and $F_2(b)$ did not match, the expected change $E[dF]$ would be of order 1, while the flow payoff and normal return terms are both of order Δt. This would violate arbitrage. If the derivatives $F_1'(b)$ and $F_2'(b)$ were unequal, then $E[dF]$ would be of order $\sqrt{\Delta t}$; this still would not be compatible with the other two terms of order Δt.

3.9. Example: temporary suspension

Return to the firm of section 3.6 just above, with two modifications. First, we remove the imposed price ceiling. Second, and more important for the current context, we now follow McDonald and Siegel (1985), and let the firm suspend operations when $P < W$.

Now the flow of profits $g(P)$ has the piecewise linear form

$$g(P) = \begin{cases} 0 & \text{if } P < W \\ P - W & \text{if } W < P. \end{cases}$$

The differential equation for $G(P)$ correspondingly has two different forms in the two regions. Let $G_1(P)$ be the solution in the region $P < W$ and $G_2(P)$ that in the region $W < P$. We have

$$G_1(P) = D_1 P^\delta, \tag{3.13}$$

and

$$G_2(P) = C_2 P^{-\gamma} + P/(\rho - \nu) - W/\rho. \tag{3.14}$$

Here we have used two limiting arguments: one as P goes to zero to eliminate the $P^{-\gamma}$ term in $G_1(P)$ and the other as P goes to ∞ to eliminate the P^δ term in $G_2(P)$.

That still leaves two undetermined constants. To determine them, we have the value matching and smooth pasting conditions of (3.12), namely

$$D_1 W^\delta = C_2 W^{-\gamma} + W/(\rho - \nu) - W/\rho,$$

$$\delta D_1 W^{\delta-1} = -\gamma C_2 W^{-\gamma-1} + 1/(\rho - \nu).$$

These are two simple linear equations for D_1 and C_2, which yield

$$D_1 = \frac{(\rho + \gamma \nu) W^{1-\delta}}{\rho (\rho - \nu)(\gamma + \delta)}, \text{ and } C_2 = \frac{(\rho - \delta \nu) W^{1+\gamma}}{\rho (\rho - \nu)(\gamma + \delta)}. \tag{3.15}$$

To find the sign of C_2, observe that

$$\psi(\rho/\nu) = \rho - \nu(\rho/\nu) - \tfrac{1}{2}\sigma^2(\rho/\nu)(\rho/\nu - 1)$$
$$= -\tfrac{1}{2}\sigma^2 \rho(\rho - \nu)/\nu^2 < 0.$$

Therefore ρ/ν lies to the right of the root δ of the fundamental quadratic, or $(\rho - \delta \nu)$ is positive. Then C_2 is positive. (A similar calculation shows that D_1 is positive; that may seem obvious until we notice that ν could be negative).

Now we can compare $G_2(P)$ in (3.14) with $G_0(P)$ in (3.6). The latter was the expected present value of the firm's profits when it was forbidden to shut down. We see how the added term in $G_2(P)$ reflects the added benefit of the ability to suspend operations. The benefit is positive even when $P > W$, that is even when the suspension option is not currently being used, because there is a positive probability that it will be invoked in the future. But the added value goes to 0 as P goes to ∞, because then shutdown becomes a remote and unlikely event.

4. OPTIMAL CONTROL AND REGULATION

An important reason for developing the calculation of expected present values subject to barriers is that such barriers are optimal policies under certain conditions. The general idea is that we can alter the motion of x at a cost, and aim to exert such control on x optimally, considering the cost of the action and its effect on the expected payoff. The appropriate policy depends on the type of choice available and the nature of the cost.

The simplest control problem involves a binary discrete choice between inaction and some specified action. If nothing is done, the process x continues according to its given probabilistic law of motion. Action amounts to stopping the process and taking some terminal payoff that depends on where we choose to stop. In many applications, the desirability of action is monotonic in x. For definiteness, suppose it is increasing. Then there is a critical value say b such that inaction is optimal so long as x remains below b, and action becomes optimal when x reaches b. Thus b serves as an absorbing barrier. It remains to characterize the optimal choice of b.

In other contexts, we let the process continue but alter its position or law of motion. If there were no cost of changing x, we would continuously keep it at the point that maximizes the flow payoff $f(x)$. When action is costly, this is not optimal. The form of the optimal policy depends in a very intuitive way on the cost of action.

If the cost of changing x is strictly convex, and the marginal cost goes to zero as the amount of the change goes to zero, then some corrective action will be taken as soon as x deviates from the level that maximizes $f(x)$. But the whole deviation will not be corrected at once, because the marginal benefit from the last little bit of adjustment is only of the second order. Thus control will be exercised gradually and continuously rather than suddenly upon reaching a threshold. Familiar quadratic adjustment costs gives rise to such gradualist policies.

In other cases, it is optimal not to exert control continuously, but only when x reaches some critical levels. Such control is sometimes called *regulation*.

If the cost of changing x has a lump-sum component, then it is not worth correcting small deviations of x from its ideal. When the deviation reaches a critically large level, say x reaches b, it is suddenly moved to c and the process recommences from this new starting point. A policy of this kind was termed *resetting* in Section 3.4. It remains to characterize the optimal values of b and c. This type of policy is sometimes called *impulse* control. Probably the best-known example is the $(S - s)$ rule in inventory management.

If the cost of changing x is proportional to the extent of the change, once again it is not worth correcting small deviations, but for a different reason. Suppose the cost is m per unit distance. So long as the flow cost of the deviation of x from its ideal level is differentiable, the marginal benefit from correcting very small deviations will be of the second order of smallness. It will be outweighed by the marginal cost m, so the small deviation will not be corrected. The marginal gain from moving x by a unit distance toward the ideal will increase as the starting point goes farther from the ideal. Ultimately the gain will exceed the cost, and some move toward the ideal will be made, but it will stop short of the ideal. We will stop at the point where the marginal gain from moving another small unit distance equals the marginal cost m. Suppose this occurs at $x = b$, and that ideal is to the left of b. Then control will be exercised to stop the x process from ever exceeding b. If it attempts to do so, it will be nudged back to b. In other words, b will become an upper reflecting barrier. Again, the aim is to characterize the optimal b. This form of policy, where the process is merely prevented from crossing a barrier by reflections when the barrier is hit, is sometimes called *barrier* or *instantaneous* control.

The above claims that certain types of policies are optimal for certain types of adjustment costs are all intuitively clear, but formal proofs can be quite difficult. I shall refer the interested reader to Malliaris and Brock (1983, pp. 124–126) or Karatzas and Shreve (1988) for the stopping problem, Flemming and Rishel (1975, Chapter 6) for the case of continuous control, Harrison, Sellke and Taylor (1983) for the lump-sum cost case, and Harrison and Taskar (1983) and Harrison (1985, Chapter 6) for the linear cost case. I shall take it for granted that thresholds or barriers are the appropriate *forms* of optimal policies, and concentrate on the methods for finding the optimum *positions* of the the thresholds or barriers.

4.1. Stopping

Suppose that, left to itself, the underlying state variable x would follow the Brownian motion (1.1), and a flow payoff $f(x)$ would accrue. But at any instant, we can stop the motion, and collect a terminal payoff $W(x)$. Some or all these payoffs could be negative (costs). The aim is to choose the stopping policy to maximize the expected present value of the payoff.

In most applications, the relative merit of stopping is monotonic in x. Without going into the details here, I shall simply assume that there is a critical value $x = b$ such that stopping is optimal to its right and continuing is optimal to its left.

We start with the solution (3.1) that gives $F(x)$ for a fixed b, and then consider the optimum choice of b. Thus

$$F(x) = F_0(x) + Ae^{-\alpha x} + Be^{\beta x},$$

where $F_0(x)$ is the expected present value of the flow $f(x)$ when we treat the x process as unrestricted. In the present instance, this amounts to ignoring the absorbing barrier at b. Then the other two terms are the additional effect of the barrier, and the constants A, B are to be determined.

We also saw that, with no lower barrier, $F(x)$ must be approximated by $F_0(x)$ as x goes to $-\infty$, for which we must set $A = 0$. Then only the coefficient B remains to be determined as a function of the critical level b. To clarify this dependence, let us write

$$F(x; b) = B(b)e^{\beta x} + F_0(x). \tag{4.1}$$

The first argument x is the state variable; the second is just a reminder that the whole function $F(x)$ shifts with changes in the parameter b. The function $B(b)$ is determined from the value matching condition (3.4), namely

$$F(b; b) \equiv B(b)e^{\beta b} + F_0(b) = W(b). \tag{4.2}$$

Note that the choice of b affects $F(x; b)$ in a very particular way. Since $e^{\beta x}$ is always positive, any change in b either raises or lowers the whole function $F(x; b)$, depending on whether $B(b)$ increases or decreases. This greatly simplifies the maximization problem. We should simply choose b to maximize $B(b)$. The first-order condition for that is $B'(b) = 0$.

Now differentiate (4.2) totally:

$$B'(b)e^{\beta b} + B(b)\beta e^{\beta b} + F_0'(b) = W'(b).$$

Substituting $B'(b) = 0$, we get

$$B(b)\beta e^{\beta b} + F_0'(b) = W'(b).$$

The left hand side is $F_x(b; b)$. If we now simplify the notation by dropping the explicit dependence on the parameter b, this becomes $F'(b)$. Thus the condition for the optimality of b is

$$F'(b) = W'(b). \tag{4.3}$$

In conjunction with the value matching condition, this says that the graphs of the functions $F(x)$ and $W(x)$ should meet tangentially, or join smoothly, at b. It is a Smooth Pasting Condition, not only in its form, but also in its substance as a requirement of optimality.

Now we can regard the value matching and the smooth pasting conditions as two equations that simultaneously define the optimal stopping point b and the constant B. Figure 4.1 illustrates the solution.

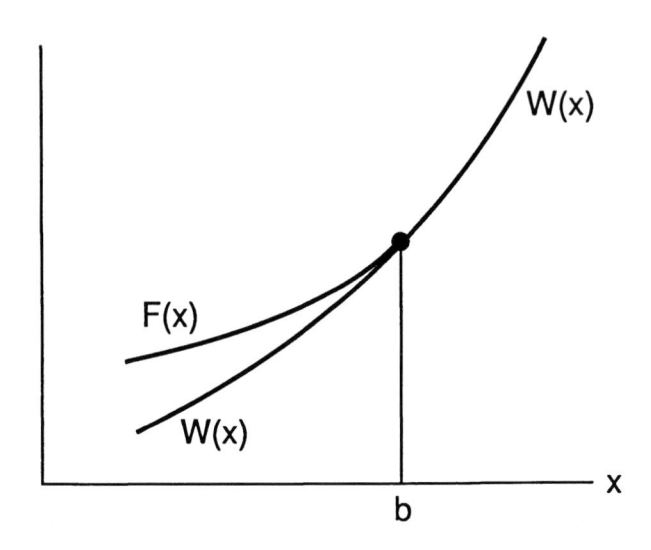

FIGURE 4.1 Smooth pasting.

We can get a better intuition for smooth pasting by using the discrete approximation to Brownian motion. At the supposed optimum b, as shown in the Figure, the slopes of $F(x)$ and $W(x)$ are equal at b. Suppose the contrary, and show that a suitably constructed alternative policy can do better. The slopes could have been unequal in either of two ways. First, we could have $W'(b) < F'(b)$. That would give $W(x) > F(x)$ to the left of b, making immediate termination optimal there, contradicting the supposed optimality of b. Or we could have $W'(b) > F'(b)$. We show that this implies a clear preference for inaction at $x = b$.

Starting at b, immediate stopping yields the terminal payoff $W(b)$. Instead, consider the following policy defined using the discrete approximation. Do not stop the process right away; wait for the next small finite time step Δt, and then reassess the situation. If the next increment is Δh, stop, and if it is $-\Delta h$, continue. Under our assumption $W'(b) > F'(b)$, this alternative policy offers a positive net gain over the supposedly optimal policy of stopping at b, because we get an average of a move to the right up a steeper curve and a move to the left along a flatter curve. More formally, we get the flow $f(b)\Delta t$ over the small time step, and the discounted value of the continuation. The total is

$$f(b)\Delta t + (1 - \rho\Delta t)(pW(b + \Delta h) + qF(b - \Delta h))$$
$$= f(b)\Delta t + (1 - \rho\Delta t)\{F(b) + [pW'(b) - qF'(b)]\Delta h + \ldots\}$$
$$= W(b) + \tfrac{1}{2}[W'(b) - F'(b)]\Delta h + \ldots$$
$$> W(b)$$

where I have used the definitions (1.5′) of p and q and the value matching condition (4.2), and have at each step retained only the leading terms of the expansion on the right hand side. (Note that Δt is of order $(\Delta h)^2$.) Thus the alternative policy does better than immediate stopping at b. This contradicts the supposed optimality of b, thereby ruling out $W'(b) > F'(b)$.

4.2. Example: irreversible investment

This follows McDonald and Siegel (1986). Consider a single discrete irreversible investment project that can be initiated at any time by incurring the sunk cost K. Once started, it will earn a cash flow of X that follows a geometric Brownian motion with parameters ν and σ. The current value of X is observable, and it guides the decision of whether to invest. Higher values of X make the investment more attractive, so there will be a critical value b such that investment is optimal to its right and inaction is optimal to its left.

The expected present value of the project launched when the current cash flow is X is easily calculated; it is $X/(\rho - \nu)$. The net payoff upon investment is therefore

$$W(X) = X/(\rho - \nu) - K.$$

No cash flow accrues while waiting, therefore for $X < b$ the expected present value has the form

$$G(X) = DX^\delta$$

in usual notation. This value arises solely from the probability of X reaching the investment trigger some time in the future.

At the optimal b, the two functions should be smoothly pasted together. Therefore

$$Db^\delta = b/(\rho - \nu) - K, \quad \delta Db^{\delta-1} = 1/(\rho - \nu).$$

Eliminating D between the two, we find

$$b = \frac{\delta}{\delta - 1} (\rho - \nu)K. \tag{4.4}$$

The simplest interpretation of this occurs when $\nu = 0$, so the cash flow process has no trend. Then ρK represents the normal return on the investment. The conventional Marshallian criterion would tell us to invest when the expected cash flow exceeds this. But the critical level in (4.4) is higher by a factor $\delta/(\delta - 1)$. This is because the option of waiting a little longer has value. For details see Pindyck (1991) and Dixit (1992 a).

4.3. Convex costs: continuous control

Suppose x follows the Brownian motion (1.1), but we can change the natural trend rate μ to $(\mu + u)$ at a cost $c(u)$ per unit time, where c is strictly convex and $c'(0) = 0$. Consider the policy of controlling u to maximize the expected present value of the flow payoff $f(x)$ net of the cost of control. Let $V(x)$ denote the optimal value starting at x. Considering the next small interval dt of time, we have

$$V(x) = \max_u [f(x)dt - c(u)dt + E[V(x + dx)e^{-\rho dt}]].$$

Using Itô's Lemma as before, this becomes

$$\rho V(x) = \max_u [f(x) - c(u) + (\mu + u)V'(x) + \tfrac{1}{2}\sigma^2 V''(x)]. \tag{4.5}$$

The first-order condition for u is

$$c'(u) = V'(x). \tag{4.6}$$

This equality of current marginal cost and future marginal benefit should be familiar, for example from dynamic programming formulations of saving. Then we can express the optimal u as a function of x, and substitute in (4.5) to get a differential equation for x. Its solution, with appropriate end-point conditions, defines $V(x)$, and that in turn allows us to find the optimal policy.

In the quadratic case $c(u) = \frac{1}{2} k u^2$, we get $u = V'(x)/k$, and then (4.5) becomes

$$\tfrac{1}{2} \sigma^2 V''(x) + \mu V'(x) - \rho V(x) + \tfrac{1}{2}[V'(x)]^2/k + f(x) = 0.$$

This is a non-linear differential equation, which needs numerical solution in most cases.

For an application to investment, see Abel (1983).

4.4. Lump-sum costs: impulse control

Suppose that by incurring a lump-sum cost k, the x process can be instantaneously reset at any desired point. Consider the case where the flow payoffs are such that very high values of x get successively less desirable. When x reaches b, we reset it to $c < b$ and continue the process.

As in the case of stopping, we write the expected present value as

$$F(x) = F_0(x) + B e^{\beta x}, \tag{4.7}$$

or showing the dependence on the barriers explicitly,

$$F(x; b, c) = F_0(x) + B(b, c) e^{\beta x},$$

where we have used the limit as x goes to $-\infty$ to set $A = 0$. Once again, changes in b or c shift the whole function $F(x; b, c)$ in the same direction for all x. Therefore the optimal choice of these parameters should maximize $B(b, c)$. The first-order conditions for this are $B_b(b, c) = 0$ and $B_c(b, c) = 0$.

The value matching condition is

$$F(b) = F(c) - k. \tag{4.8}$$

Introducing the dependence on the parameters explicitly, this is

$$F(b; b, c) = F(c; b, c) - k,$$

or

$$B(b,c)[e^{\beta b} - e^{\beta c}] = F_0(c) - F_0(b) - k.$$

Differentiating with respect to b gives

$$B_b(b,c)[e^{\beta b} - e^{\beta c}] + B(b,c)\beta e^{\beta b} = -F_0'(b).$$

Using the first-order condition $B_b(b,c) = 0$, we get

$$F_0'(b) + B(b,c)\beta e^{\beta b} = 0,$$

or $F_x(b;b,c) = 0$. Similar calculation for c gives $F_x(c;b,c) = 0$.

Resorting to the original notation where b and c were not shown as explicit arguments in F, these conditions are simply

$$F'(b) = 0 = F'(c). \tag{4.9}$$

These are smooth-pasting type conditions at both the starting point and the end point of the optimal resetting. Now we can regard the value matching condition in (4.8) and the two smooth pasting conditions in (4.9) as three equations that define the optimal policies b, c and the undetermined constant B.

In a two-sided impulse control problem, excessively low values of x are also undesirable, and we have to find three critical numbers $a < c < b$. The process is allowed to continue in the interval (a, b), but when it hits either end point, it is reset to c. Now we cannot use a limiting argument as x goes to $-\infty$, and must leave $F(x)$ with two undetermined constants:

$$F(x) = F_0(x) + Ae^{-\alpha x} + Be^{\beta x} \tag{4.10}$$

We now have two value matching conditions

$$F(a) = F(c) - k, \quad F(b) = F(c) - k. \tag{4.11}$$

Once again, the parameters affect F only through positive multiples of A and B, so the aim is to make A and B as large as possible. Moreover, each of the parameters moves A and B in the same direction, so there is no conflict between the maximization of the two. To see this, take the value matching condition linking b and c and differentiate it with respect to a. This yields

$$A_a e^{-\alpha b} + B_a e^{\beta b} = A_a e^{-\alpha c} + B_a e^{\beta c},$$

or

$$A_a[e^{-\alpha b} - e^{-\alpha c}] + B_a[e^{\beta b} - e^{\beta c}] = 0.$$

THE ART OF SMOOTH PASTING   41

Since α, β are positive, and $c < b$, the two expressions in the two square brackets have opposite signs. Therefore A_a and B_a have the same sign. Similarly for the derivatives with respect to b and c.

It remains to maximize A and B. Following familiar steps, that yields three smooth pasting conditions

$$F'(a) = F'(c) = F'(b) = 0. \qquad (4.12)$$

The two value matching conditions in (4.11) and the three smooth pasting conditions in (4.12) comprise five equations that determine the two constants A, B and the three optimal policy parameters a, b, c.

4.5. Example: menu costs

This is based on Dixit (1991 a). Suppose x follows the trendless process $dx = \sigma dw$. Suppose the flow cost is x^2, and the lump-sum cost of resetting x is k. The policy aims to minimize the total expected cost.

Now $F_0(x)$ is a special case of (2.6) above (with $\mu = 0$). The quadratic that determines α and β is

$$\phi(\xi) = \rho - \tfrac{1}{2}\sigma^2 \xi^2 = 0.$$

Therefore $\beta = \alpha = \sqrt{2\rho}/\sigma$. Then

$$F(x) = \sigma^2/\rho^2 + x^2/\rho + Ae^{-\alpha x} + Be^{\alpha x}.$$

The symmetry of the problem quickly gives $A = B$ and $a = -b$, $c = 0$. The value matching condition becomes

$$A[e^{\alpha b} + e^{-\alpha b}] + b^2/\rho = k + 2A,$$

and the smooth pasting condition is

$$\alpha A[e^{\alpha b} - e^{-\alpha b}] + 2b/\rho = 0.$$

Eliminating A, we have an equation that defines b:

$$-[e^{\alpha b} + e^{-\alpha b} - 2]/[e^{\alpha b} - e^{-\alpha b}] = \alpha[k - b^2/\rho]/(2b/\rho).$$

When k is small, this has the approximate solution

$$b = (6k\sigma^2)^{1/4}. \qquad (4.13)$$

Note that if $k \sim \varepsilon^4$ for small ε, then $b \sim \varepsilon$. In other words, *fourth-order small menu costs have first-order effects*.

4.6. Linear costs: barrier control

Now the cost of changing x is proportional to the distance through which it is moved, say m per unit. As before, begin with the case when high values of x become increasingly less desirable. Now we set up a reflecting upper barrier b, and write

$$F(x) = F_0(x) + Be^{\beta x}, \tag{4.14}$$

where B is determined using the condition appropriate to the barrier. In the discussion following (3.5), we saw this to be the condition

$$F'(b) = -m. \tag{4.15}$$

This being a condition on a first-order derivative is sometimes called a smooth pasting condition, but here it holds for any exogenously specified b and has no optimizing role. Using it on (4.14), we have

$$F_0'(b) + \beta B e^{\beta b} = -m.$$

To find the optimum (B-maximizing) choice of b, differentiate this again:

$$F_0''(b) + \beta B_b e^{\beta b} + \beta^2 B e^{\beta b} = 0.$$

Use the first-order condition $B_b = 0$ to get

$$F_0''(b) + \beta^2 B e^{\beta b} = 0,$$

which is just

$$F''(b) = 0. \tag{4.16}$$

The optimality condition now pertains to the second-order derivative of F; it is sometimes called the Super Contact Condition.

In the resetting or impulse control problem, at the optimal choice of the barrier we had a smooth pasting condition, involving the first-order derivative of $F(x)$. By contrast, the condition at any exogenously specified and not necessarily optimal barrier pertained to the values of the functions, which might be called a 'zeroeth order derivative' condition. For the barrier control policy in the case of linear costs, the order of differentiation shifts up by one: the first order or smooth pasting condition holds for any position of the barrier, optimal or otherwise, and the equality of second order derivatives or super contact characterizes the optimal barrier. For more details, see Dumas (1991).

Once again, we regard (4.15) and (4.16) as two equations that define B and b, completing the solution. The extension to two-sided regulation follows familiar steps.

4.7. Some geometry and intuition

Now consider a case that combines those of Sections 4.4 and 4.6, allowing for lump-sum as well as linear costs. Suppose the cost of moving x upward has a lump-sum component K_u and a linear component proportional to m_u. Similarly K_d and m_d for downward movements. Lump-sum costs imply a range of inaction, and sudden jumps when the extremes of this range are reached, as in Section 4.4. But because there is a linear component of the cost, the jumps aim at points where the marginal benefits of further change fall to the level of the marginal cost. Therefore upward and downward jumps do not have a common terminal point. The optimal policy is characterized by four threshold levels of x, say a, c, d and b in increasing order. Inaction is optimal when $a < x < b$. When x falls to a, it is instantly reset to c; when it rises to b, it is reset to d. Following the same steps as in Sections 4.4 and 4.6, we get the value matching conditions

$$F(c) - F(a) = K_u + m_u(c - a),$$
$$F(d) - F(b) = K_d + m_d(b - d), \qquad (4.17)$$

and the smooth pasting conditions

$$F'(c) = F'(a) = m_u, \quad F'(d) = F'(b) = -m_d. \qquad (4.18)$$

A simple figure, based on Constantinides and Richard (1978) and Harrison, Sellke and Taylor (1983) illustrates this solution, and helps develop some useful intuition. We use $F'(x)$, the marginal benefit of being able to start at a slightly higher value of x. Differentiating the general solution (4.10), we write

$$F'(x) = F_0'(x) - \alpha A e^{-\alpha x} + \beta B e^{\beta x}. \qquad (4.19)$$

If the flow payoff function $f(x)$ is concave, as befits a maximization problem, then $F_0(x)$ inherits this concavity. Then $F_0'(x)$ is a decreasing function. Since the last two terms in (4.10) are the addition to the expected present value made possible by the ability to control the x process, the constants A and B are positive. Then in (4.19), $-\alpha A e^{-\alpha x}$ is negative and increasing, while $\beta B e^{\beta x}$ is positive and increasing. The former dominates as x goes to $-\infty$ and the latter as x goes to ∞. The remaining term in (4.19), $F_0'(x)$, is relatively more important for middle values of x. Putting all this information together, the graph of $F'(x)$ is as shown in Figure 4.2.

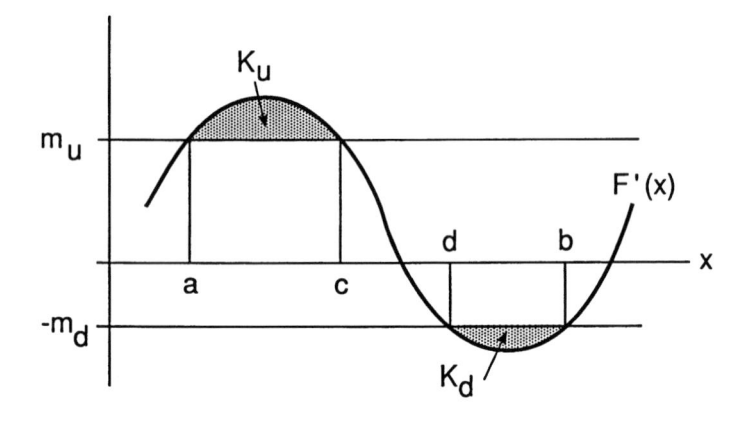

FIGURE 4.2 Optimal regulation.

The figure also shows two horizontal lines, at heights m_u and $-m_d$. Let the former cut $F'(x)$ at a, c and the latter at d, b as shown. We can move $F'(x)$ around by changing the constants A, B. This is to be done until the shaded areas trapped between the curve and the horizontal lines, respectively, equal K_u and K_d as shown. Then a, c, d and b are the thresholds that define the optimal policy described above.

To prove the assertion of optimality, let us verify that the appropriate conditions are met. For upward movements, the height of the $F'(x)$ curve at both a and c equals m_u; these are the smooth pasting conditions for upward moves in (4.18). In words, they say that at both the initial and the terminal points of the sudden jump in x the marginal benefit equals the marginal cost. Next, the total area under the $F'(x)$ curve between a and c is just $F(c) - F(a)$. Subtracting the rectangle $(c - a)m_u$ leaves the shaded area, and that is equal K_u. This is just the upward value matching condition in (4.17). In words, it says that the total benefit form the jump should equal its total cost. The downward conditions are verified similarly.

We can now recover the earlier special cases of purely lump-sum and purely linear costs. If the cost is purely lump-sum, $m_d = 0 = m_u$ and the points c and d coalesce. This is the three-point policy of section 4.4 above. If the downward cost has no lump-sum component, $K_d = 0$, the shaded area vanishes, and b merges with d. At this point $F'(x)$ becomes horizontal, so $F''(b) = 0$, which is just the super contact condition (4.16). Similarly, $F''(a) = 0$ if $K_u = 0$. For more details see Dixit (1991 b).

4.8. Example: competitive industry

Consider an industry where each unit of capital costs K to install, and is irreversibly committed thereafter. Each installed unit of capital produces a unit of output at zero variable cost. If Q units of output are being produced, the benefit (consumer utility) is $XS(Q)$, where $S(Q)$ is the usual increasing concave function, and X follows a geometric Brownian motion with parameters ν and σ. The aim is to maximize the expected present value of utility, net of the costs of installing new capital.

No new investment is made while X stays below a threshold level that is itself a function of the current Q. Our aim is to find this threshold. As with a discrete project of section (4.2), we begin with the region where no investment takes place, so Q is constant.

Now the problem has two state variables, the current X and the currently installed capacity which equals the current output Q. Define $F(X, Q)$ as the expected present value. While Q is held constant, the standard arbitrage equation gives

$$\tfrac{1}{2}\sigma^2 X^2 F_{xx}(X, Q) + \nu X F_x(X, Q) - \rho F(X, Q) + XS(Q) = 0. \tag{4.20}$$

This is a partial differential equation, but for each fixed Q we can treat it as an ordinary one with X as the independent variable, provided we remember that any constants in the solution will actually be functions of Q. Thus the general solution is

$$F(X, Q) = D(Q)X^\delta + XS(Q)/(\rho - \nu), \tag{4.21}$$

where the function $D(Q)$ remains to be determined. Once again I have used limiting arguments for small X to eliminate the negative power in the complementary function.

This is a somewhat more difficult problem because the controlled variable Q is not the same as the Brownian motion variable X. Sometimes a suitable composite variable can be used instead, but it is easier to use the intuition of the one variable case to give a heuristic solution of the two variable problem.

Since the cost of changing Q is linear, control will not be exercised continuously. If X becomes too large relative to the current Q, just enough new capacity will be installed to bring the relation between X and Q back to the threshold level. At the margin where some new capacity is being added, the marginal benefit will equal the marginal cost, that is, $F_Q(X, Q) = K$ or

$$D'(Q)X^\delta + XS'(Q)/(\rho - \nu) = K. \tag{4.22}$$

The super contact condition will be $F_{QQ}(X, Q) = 0$. By differentiating $F_Q(X, Q) = K$ totally, this can equivalently be written as $F_{XQ}(X, Q) = 0$, or

$$\delta D'(Q)X^{\delta-1} + S'(Q)/(\rho - \nu) = 0.$$

Combining this with (4.22), we get the relation between X and Q that defines the margin where new investment occurs:

$$XS'(Q) = \frac{\delta}{\delta - 1}(\rho - \nu)K. \tag{4.23}$$

Note that the left hand side is just the current price (marginal utility). Thus we have a striking parallel between (4.23) for industry investment and (4.4) for a single firm or project. In either case, investment is not undertaken unless the current return is at a premium $\delta/(\delta - 1)$ over the normal return.

For an alternative insight, think of optimization problem as being solved in a decentralized manner, namely a rational-expectations equilibrium of price-taking risk-neutral firms, each of which can become active by installing a unit of capacity at cost K. Competitive entry ensures that the output price P never rises above the level given by the right hand side of (4.23). Each firm makes its entry decision knowing that the price process has this upper reflecting barrier. This cutoff of the upside price potential turns out to reduce the value of immediate investment and that of waiting alike, leaving the firm's entry threshold price the same as in (4.4). Then each firm's entry threshold coincides with the price ceiling that was assumed by each in its entry calculation as arising from the decisions of the others. This 'fixed point' argument completes the determination of competitive equilibrium. For more details, see Dixit and Pindyck (1992, Chapter 8).

5. GENERALIZATIONS

The set of techniques developed and applied above are already rich enough for many uses, but some further generalizations are available. I shall mention just two. One is to processes more general than Brownian motion, the other to finite-horizon problems.

5.1. Mean-reverting processes

In economic applications, there is often reason to believe that general equilibrium considerations will give rise to mean-reversion in the basic stochastic process for x. This alters the basic differential equation and methods for its solution, but not the fundamental conditions of value matching and smooth pasting. I shall illustrate this by reworking the menu cost model when x follows an Ornstein–Uhlenbeck process:

$$dx = -\lambda x dt + \sigma dw. \tag{5.1}$$

Let $C(x)$ denote the minimized expected discounted cost when the flow cost is x^2 and the cost of resetting is k. By symmetry we know that there will be no resetting within an interval $(-h, h)$, and when x reaches b, it will be reset to zero. In the interval $(-h, h)$, familiar steps show that $C(x)$ satisfies the differential equation

$$\tfrac{1}{2}\sigma^2 C''(x) - \lambda x C'(x) - \rho C(x) + x^2 = 0. \tag{5.2}$$

Trying a particular solution $C(x) = a + bx$ and comparing coefficients, we find

$$a = \sigma^2/[\rho(2\lambda + \rho)], \qquad b = 1/(2\lambda + \rho).$$

For the homogeneous part of the equation, try a power series expansion

$$C(x) = \sum_{n=0}^{\infty} c_n x^{2n}$$

where I have used symmetry to retain only the even powers. We get

$$0 = \tfrac{1}{2}\sigma^2 \sum_{n=1}^{\infty} c_n 2n(2n-1)x^{2n-2} - \sum_{n=0}^{\infty} c_n[2n\lambda + \rho]x^{2n}$$

$$= \sum_{n=0}^{\infty} [\sigma^2(n+1)(2n+1)c_{n+1} - (2n\lambda + \rho)c_n]x^{2n}$$

For a solution, coefficients of all powers of x must be zero. Now define the sequences $\{d_n\}$ and $\{s_n\}$ by

$$d_n = (2n\lambda + \rho)/[\sigma^2(n+1)(2n+1)] \tag{5.3}$$

and

$$s_0 = 1, \quad s_{n+1} = d_n s_n. \tag{5.4}$$

Then $c_n = s_n c_0$ for the solution.

Thus the general solution is

$$C(x) = \frac{\sigma^2}{\rho(2\lambda + \rho)} + \frac{x^2}{2\lambda + \rho} + c_0 \sum_{n=0}^{\infty} s_n x^{2n}, \tag{5.5}$$

where c_0 remains as the only undetermined constant. For a given barrier h, it is determined by the value matching condition

$$C(h) = C(0) + k,$$

or

$$\frac{h^2}{2\lambda + \rho} + c_0 \sum_{n=1}^{\infty} s_n h^{2n} = k. \tag{5.6}$$

The optimum h is then characterized by the smooth pasting condition

$$C'(h) = 0,$$

or

$$\frac{2h}{2\lambda + \rho} + 2c_0 \sum_{n=1}^{\infty} n s_n h^{2n-1} = 0. \tag{5.7}$$

After some algebra, this simplifies to

$$6\sigma^2 k/h^4 = \frac{1 + 2(s_3/s_2)h^2 + 3(s_4/s_2)h^4 + \cdots}{1 + 2(s_2/s_1)h^2 + 3(s_3/s_1)h^4 + \cdots}$$

Note that when k is small, the solution is approximately $h = (6\sigma^2 k)^{1/4}$, the same as in the case without mean reversion, namely $\lambda = 0$. Else it can be calculated numerically. Here is a sample for parameters $k = 0.2$, $\rho = 0.05$ and $\sigma = 0.1$:

λ	h
0.00	0.340
0.10	0.356
0.20	0.378
0.50	0.480
1.00	0.652
2.00	0.915

Note that the range of no action is wider when mean reversion is stronger. The fact that the process is trended toward the ideal point of its own accord increases the controller's reluctance to incur to lump-sum cost. This fits well with intuition. But there is a countervailing effect, namely the mean reversion renders the process stationary and therefore less volatile, thereby decreasing the value of the option to wait. And in some other examples the net effect goes the other way.

For more details in the context of an application to exchange rates, see Gerlach (1991).

If a geometric Brownian motion X is mean-reverting, for example

$$dX/X = -\lambda(X - X^*)dt + \sigma dw, \tag{5.8}$$

then the expected present value function $G(x)$ will satisfy

$$\tfrac{1}{2}\sigma^2 X^2 G''(X) - \lambda X(X - X^*)G'(X) - \rho G(X) + g(X) = 0. \tag{5.9}$$

The solution to this can be developed using power series. But the equation can be transformed into a standard form in mathematical physics, and the solution expressed in terms of confluent hypergeometric functions. For an example, see Bartolini and Dixit (1991).

5.2 Finite horizon

If the control problem has a finite horizon, the option to wait approaches expiry and therefore becomes valueless as we approach the horizon. Technically, time must enter as an explicit state variable and the differential equation for the value function becomes a parabolic partial differential equation. It generally needs a numerical solution. Here is a simple example; see Pindyck (1991) and the references cited there for several others.

Consider a machine that cost k (sunk) to install and m (flow) to operate. It produces a flow of output worth P, which flows the geometric Brownian motion (1.8),

$$dP/P = \nu dt + \sigma dw.$$

The machine lasts T years. If it is abandoned before its physical life is up, it will 'rust' and cannot be reused. The problem is to find the optimal abandonment policy for a machine that has y years of physical life left.

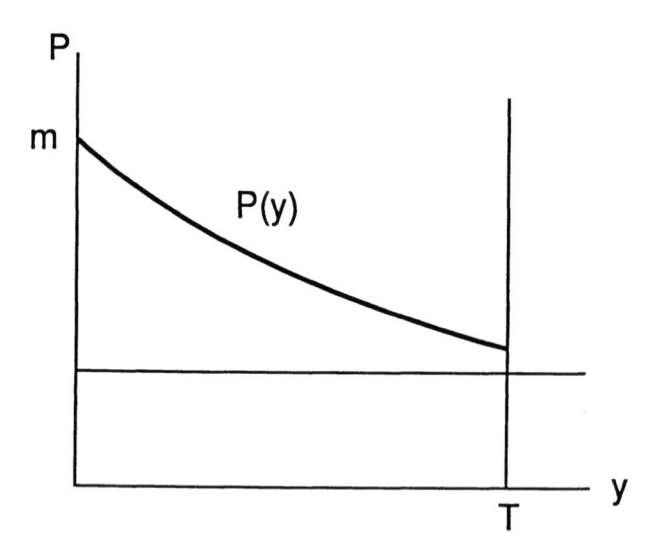

FIGURE 5.1 Depreciation and abandonment.

Let $V(P, y)$ denote the expected present value of profits starting at state (P, y). While the machine is in operation, this will satisfy

$$V(P, y) = (P - m)dt + E[V(P + dP, y - dt)e^{-\rho dt}].$$

Using Itô's Lemma in its time-dependent form (1.7), we have the differential equation

$$\tfrac{1}{2}\sigma^2 P^2 V_{PP}(P,y) + \nu P V_P(P,y) - \rho V(P,y) - V_y(P,y)$$
$$+ (P - m) = 0. \qquad (5.10)$$

The abandonment barrier is now a curve in (P,y) space, say $P = P(y)$. At all points of it, V should satisfy the value matching and smooth pasting conditions

$$V(P,y) = 0, \qquad V_P(P,y) = 0. \qquad (5.11)$$

The problem of determining such a 'free boundary' is particularly tricky, and numerical methods for it are largely *ad hoc*. In this case intuition suggests the form of the solution; which can then be computed numerically. A typical outcome is shown in Figure 5.1. When y is very small, that is, very little physical life is left, the machine will be abandoned as soon as the flow profit $(P - m)$ becomes slightly negative. The option to wait hoping for an upturn in price is of little value when the machine nears the end of its physical life. For larger y, it will be desirable to tolerate some current loss to keep alive the option of later profitable use. If T is very large, then for very large y the abandonment trigger $P(y)$ falls to the value for the infinite-time case.

6. SOME CHARACTERIZATION OF OPTIMAL PATHS

In Chapter 4 we determined optimal policies in state space; the various thresholds that governed stopping, resetting or regulation were all expressed in terms of the underlying state variable. But we are often interested in other, more explicitly dynamic, properties of a system that evolves given such policies. For example, if some action is taken when the Brownian motion of the state variable reaches a given threshold, what is the expected time until this happens? In the long run, will the system settle into a stochastically stationary state, and if so, what are the properties of the long run distribution of the state? In this chapter I illustrate some techniques for answering such questions. The chosen settings cover some important economic applications. But I do not develop the theory at a general level; that would take far too long. I merely refer the reader to books on stochastic processes for more detailed and rigorous treatments.

For sake of brevity again, I proceed on the assumption that the state variable follows the absolute Brownian motion (1.1) when no control is exercised. Geometric Brownian motion can be reduced to this case by taking logarithms with due care about Itô's Lemma; this is left as an exercise for the readers. More complicated processes such as the mean-reverting case can be handled using the methods sketched in Chapter 5 combined with numerical methods.

The general notation is as follows. The state variable has a lower barrier a and an upper barrier b. In different contexts these may be reflecting or absorbing as stated; in some cases there may be only one barrier, when we set $a = -\infty$ or $b = \infty$ as appropriate. Between these barriers, x_t follows an absolute Brownian motion with parameters μ and σ. The initial point is $x_0 = x$.

6.1 Short run: time until first action

Starting at x in the interior of the range (a, b) between the barriers, the Brownian particle will almost surely stay in the interior for a sufficiently short initial interval of time. If both barriers are finite, then after a sufficiently long interval of time it is sure to hit one or the other barrier. But if one barrier is at infinity and the trend of the Brownian motion is in that direction, then the particle may drift off in that direction and never hit the other barrier.

If the barriers are absorbing, the story ends with the first hit at a barrier. If the barrier first hit is reflecting, the processes continues, and the particle may go on to hit the other barrier or the same one again, but the event of the first hit is independent interest anyway. In this section we examine some properties of such first hits, in particular (1) the expected first time, if ever, that the particle hits either barrier or a particular barrier, and (2) the probability that the upper (resp. lower) barrier is the one reached first.

In the theory of the simple random walk, such problems are studied under the heading of 'gambler's ruin.' Imagine two players, A and B, engaged in an indefinite repetition of a game of chance. Each individual game has a dollar at stake: A pays B with probability p, and B pays A with probability $q = 1 - p$. Then the random variable x_t, defined as the cumulative winnings of B at time t, follows a random walk. This has barriers where one or the other player runs out of money. The upper barrier equals A's capital, the lower equals *minus* B's capital. We are interested in the probabilities of ruin for the two

THE ART OF SMOOTH PASTING

players, expected time to ruin, etc. See Feller (1968, pp. 344–349), and Cox and Miller (1965, Chapter 2) for a detailed development.

We can get the corresponding theory for Brownian motion by taking limits of the random walk as in Section 1.1 above. Label the discrete points by the index i. Let $x = a$ correspond to $i = 0$, and $x = b$ to $i = m$. Thus $b = a + m\Delta h$. The general point i corresponds to $x = a + i\Delta h$. This follows Cox and Miller (1965, Chapter 5).

Probabilities

Let Q_i denote the probability that the first barrier hit is a when the starting point is i. For $i = 1, 2, \ldots, m - 1$, consider the first step: this could be to $(i + 1)$ with probability p, and to $(i - 1)$ with probability q. The probabilities of hitting a first conditional on the new starting points are Q_{i+1} and Q_{i-1} respectively. Therefore we have the recurrence relation

$$Q_i = qQ_{i-1} + pQ_{i+1}. \tag{6.1}$$

This is a simple linear difference equation for the unknowns Q_i. Try a solution of the form $Q_i = \beta^i$. Substituting, β must satisfy

$$1 = q/\beta + p\beta.$$

This has solution $\beta = 1$ and $\beta = q/p$. If $p \neq q$ the two solutions are independent, and the general solution of (6.1) is

$$Q_i = A + B(q/p)^i,$$

Where A and B are constants to be determined. For that we have the end-point conditions at the barriers. If $i = 0$ we start at a and therefore $Q_0 = 1$. Similarly $Q_m = 0$. Therefore

$$A + B = 1 \quad A + B(q/p)^m = 0,$$

and

$$Q_i = \frac{(q/p)^i - (q/p)^m}{1 - (q/p)^m}. \tag{6.2}$$

The result makes good intuitive sense. First, note that Q_i decreases with i: as the initial point moves farther from a, reaching a before b becomes less likely. Next, Q_i is an increasing function of q/p: for a given initial point, a higher probability of moving down at each step implies a higher probability of reaching the lower barrier first.

The above solution becomes indeterminate (0/0) if $q = p = \frac{1}{2}$, but letting $(q/p) = 1 + \varepsilon$ and taking limits as ε goes to zero, we find

$$Q_i = 1 - i/m \qquad (6.2')$$

The same result can be obtained by a direct argument.

For Brownian motion, take $i = (x - a)/\Delta h$, $m = (b - a)/\Delta h$, and q, p defined in (1.5') above, and let $\Delta h \to 0$. We get

$$\left(\frac{q}{p}\right)^i = \left[\frac{\frac{1}{2}[1 - (\mu/\sigma^2)\Delta h]}{\frac{1}{2}[1 + (\mu/\sigma^2)\Delta h]}\right]^{(x-a)/\Delta h}$$
$$\to \exp[-(x-a)\mu/\sigma^2]/\exp[(x-a)\mu/\sigma^2]$$
$$= \exp[2(a-x)\mu/\sigma^2]. \qquad (6.3)$$

Similarly for the other term. Then the probability $Q(x)$ of reaching a first starting from x when $\mu \neq 0$ is

$$Q(x) = \frac{\exp(-2x\mu/\sigma^2) - \exp(-2b\mu/\sigma^2)}{\exp(-2a\mu/\sigma^2) - \exp(-2b\mu/\sigma^2)}. \qquad (6.4)$$

When $\mu = 0$, a limiting argument gives

$$Q(x) = (b - x)/(b - a). \qquad (6.4')$$

The formula for the probability $P(x)$ of reaching b before a can be found similarly. Note that $P(x) + Q(x) \equiv 1$; there is zero probability that the particle wanders between two finite barriers without ever hitting either.

The case of only one barrier, say at $x = a$, can be found by letting b go to ∞ in (6.4). We have

$$Q(x) = \begin{cases} 1 & \text{if } \mu \leq 0 \\ \exp[2(a-x)\mu/\sigma^2] & \text{if } \mu > 0. \end{cases} \qquad (6.5)$$

Note that $Q(x) < 1$ if $\mu > 0$; if the trend is away from the barrier then there is positive probability that the particle drifts away without ever hitting the barrier.

Expected times

When we consider the time at which the particle will first reach one specified barrier, we must be careful to specify whether the other barrier is reflecting or absorbing. For example, the event of hitting a will be importantly affected by whether the motion will be stopped in the event b is reached first.

Our analysis begins, however, with an interior initial point. For the first step from such a point, considerations of repeated hits are irrelevant and purely local relationships hold true no matter what the nature of the barriers. This leads to a recurrence relation, or difference equation, whose general solution can be written in terms of some unknown parameters. These are determined by conditions at the barriers, and that is where the exact specification of the barrier type comes into play.

Therefore let T_i denote the expected time to any specified future event starting at i. For $0 < i < m$, the first step requires the unit of time Δt. The remaining expected time is computed starting at the new points $(i + 1)$ or $(i - 1)$ with their respective probabilities p, q. Therefore we have the recursion

$$T_i = \Delta t + qT_{i-1} + pT_{i+1}. \tag{6.6}$$

Note that some events may have infinite expected time, either because there is a positive probability that they never happen, or because probabilities of very long delays decrease too slowly. In such cases this equation takes the form $\infty = \infty$, which is true but uninformative. For now I shall proceed assuming that T_i is finite, and examine the issue in specific cases as they arise.

When $q \neq p$, the difference equation (6.6) has the general solution

$$T_i = -i\Delta t/(p - q) + A + B(q/p)^i.$$

In the Brownian motion limit, we use the expression (6.3) for the limit of $(q/p)^i$ to write this as

$$T(x) = -(x - a)/\mu + A + B\exp[2(a - x)\mu/\sigma^2],$$

assuming $\mu \neq 0$. The end-point conditions that determine the constants A, B depend on the exact specification of the problem. Let us consider a couple of examples.

First consider two barriers at a and b, and let T_i be the expected first time to reaching either barrier. In effect, both barriers are treated as absorbing. Then $T_0 = 0 = T_m$. Solving for A and B, we find

$$T_i = \frac{\Delta t}{p - q}\left(m\frac{1 - (q/p)^i}{1 - (q/p)^m} - i\right). \tag{6.7}$$

If $q = p = \frac{1}{2}$, we get, either taking a limit or using a direct argument,
$$T_i = i(m - i)\Delta t. \tag{6.7'}$$
The corresponding formulas for Brownian motion are: when $\mu \neq 0$,
$$T(x) = \frac{b - a}{\mu}\left[\frac{\exp(-2a\mu/\sigma^2) - \exp(-2x\mu/\sigma^2)}{\exp(-2a\mu/\sigma^2) - \exp(-2b\mu/\sigma^2)} - \frac{x - a}{b - a}\right], \tag{6.8}$$
and for $\mu = 0$
$$T(x) = (b - x)(x - a)/\sigma^2. \tag{6.8'}$$
If there is only one barrier at $x = a$, we let b go to ∞ and find
$$T_i = \begin{cases} \infty & \text{if } p \geq q \\ i\Delta t/(q - p) & \text{if } p < q. \end{cases} \tag{6.9}$$
For Brownian motion, we have the limits
$$T(x) = \begin{cases} \infty & \text{if } \mu \geq 0 \\ -(x - a)/\mu & \text{if } \mu < 0. \end{cases} \tag{6.10}$$

Note that the expected time is infinite when the drift is away from the barrier, and also when there is no drift at all. The former is intuitive, but the latter is more subtle. As we saw above, the probability of an eventual hit is unity. But very long excursions away from the barrier can occur. Then the probabilities do not fall sufficiently fast for successively longer hitting times, and the expectation, which is the sum of the products of the times and their probabilities, diverges.

Now consider another case, where T_i is the expected time to reaching the particular barrier at a, while the barrier at b is reflecting. Thus, between the starting time and the first time to reaching a, we allow the possibility of any number of reflections at b. Again we have $T_0 = 0$ for the discrete random walk, or $T(a) = 0$ in the Brownian limit. But the boundary condition at b needs more care. If the starting point is m, the next step is sure to be downward. Therefore
$$T_m = \Delta t + T_{m-1}.$$
In the Brownian motion limit, this becomes
$$-T'(b)\Delta h + \ldots = \Delta t.$$

But $\Delta t/\Delta h$ goes to zero, so we must have $T'(b) = 0$, a 'smooth pasting' like condition at a reflecting barrier. Using these conditions, we find, when $\mu \neq 0$,

$$T(x) = -\frac{x-a}{\mu} + \frac{\sigma^2}{2\mu^2}\left[e^{2(b-a)\mu/\sigma^2} - e^{2(b-x)\mu/\sigma^2}\right], \quad (6.11)$$

and when $\mu = 0$,

$$T(x) = (x-a)(2b-x-a)/\sigma^2. \quad (6.11')$$

Other combinations of barriers can be examined similarly as needed in specific applications. One can take this analysis further and find the whole distribution of the first passage times for random walks and Brownian motion. I shall leave this to more specialized books, for example Cox and Miller (1965, Sections 2.2, 5.10), Bhattacharya and Waymire (1989, Chapter 1 especially sections 3, 4, 9, 10), Feller (1968, pp. 349-359), Harrison (1985, pp. 38-44), Karlin and Taylor (1975, pp. 345-355).

Example: Menu Costs

The model and the notation is the same as in Section 4.5. Suppose x follows a Brownian motion without drift ($dx = \sigma dw$) between barriers symmetrically placed around zero, say at $\pm b$. Whenever either barrier is hit, the particle is instantaneously moved to the central point 0, and the motion restarted.

From a general initial point x, the expected time to the first such resetting is found using (6.8') above:

$$T(x) = (b^2 - x^2)/\sigma^2.$$

Now suppose b is chosen optimally balance quadratic flow costs x^2 of deviation of x from the origin, and lump sum costs k of resetting. For small k, equation (4.13) gave the approximate formula

$$b = (6k\sigma^2)^{1/4}.$$

Using this, we get

$$T(x) = \sqrt{6k}/\sigma - x^2/\sigma^2.$$

Consider what happens to $T(x)$ if σ increases and the threshold b is changed to its new higher optimal value. Two offsetting influences arise. A given Brownian motion should on average take longer to reach the new wider thresholds. But the motion itself has become more volatile, and should on average reach a given threshold sooner. Both these effects are present in our model. The balance depends on the initial position x. If $x = b$ in the original situation, then $T(x)$ is initially zero and can only increase as b increases in response to the rise in σ. But if $x = 0$ initially, our formula shows that $T(0)$ decreases as σ increases. The increase in b is less than proportionate to that in σ, so from the origin the new threshold is reached sooner on average.

We see that an increase in the underlying uncertainty can have very different effects on thresholds in the state space and on the time taken to reach them. Therefore great care should be exercised when formulating hypotheses to test such models. For more on this, see Dixit (1992 b).

6.2. Long run: stationary distribution and average action

Suppose the state variable x starts at x_0 and follows the Brownian motion (1.1). If no control is ever exercised, we know that at a future time t the distribution of x_t is normal with mean $x_0 + \mu t$ and variance $\sigma^2 t$. As t goes to infinity, this does not converge to a stationary long-run distribution. But when the motion is subject to barriers or thresholds imposed by the control policies, a long-run stationary distribution may exist, and it can give us some useful economic information. For example, if employment fluctuates between two reflecting barriers because of hiring and firing costs, we want to know the long-run average level of employment; see Bentolila and Bertola (1990). Here I drive some such formulas using the discrete approximation method.

Let Π_i be the stationary probability of state i, that is the proportion of time the particle spends at i in the long run. We get a recursion by looking to the predecessors of i. If i lies in the interior of the range, $(1 < i < m - 1)$, then the particle comes to i from $(i + 1)$ with probability q or from $(i - 1)$ with probability p. Therefore

$$\Pi_i = q\Pi_{i+1} + p\Pi_{i-1}. \tag{6.12}$$

Think of this as an equation of material balance for a whole mass of particles. Of the particles occupying the position i at a given instant, at the next step of the random walk a fraction p will move to the right to $(i + 1)$ and a fraction q will move to the left to $(i - 1)$. Then in equation (6.12) the left hand side is the total outflow from i, and the right hand side the total inflow to i, that occur during this step. For stationary the two flows should be equal. The general solution of (6.12) follows by familiar methods. We have

$$\Pi_i = \begin{cases} A + B(p/q)^i & \text{if } p \neq q \\ A + Bi & \text{if } p = q = \tfrac{1}{2}. \end{cases} \quad (6.13)$$

In the Brownian motion limit,

$$\Pi(x) = \begin{cases} A + B \exp(2(x - a)\mu/\sigma^2) & \text{if } \mu \neq 0 \\ A + Bx & \text{if } \mu = 0. \end{cases} \quad (6.14)$$

The constants A and B are to be determined in each case, and may have different values in the different cases.

As with expected hitting times, the argument that led to the difference equation (6.12) and its general solution (6.13) is purely local. It holds at any interior point of the range, independently of the nature of barriers. What the barriers do is to provide boundary conditions that determine the constants A and B in the general solution.

The case of absorbing barriers is of little interest; in the long run the whole probability will pile up at the barriers unless the drift of the motion is away from it. I shall offer two illustrations that have applicability; one for reflecting barriers and one for resetting.

Reflecting barriers

I shall interpret reflecting barriers in a particular way that is suited to the calculations below. I shall suppose that the particle is reflected instantaneously upon hitting the barrier. Thus, starting at 1, the particle either moves up to 2 with probability p, or moves down to 0 and is immediately reflected back to 1 with probability q. Similarly from $(m - 1)$ adjacent to the upper barrier at m. Thus the actual barrier locations 0 and m are never occupied for finite time. The alternative would be to allow occupancy of the actual barrier locations for a unit period before reflection. The two approaches yield the same ultimate results, especially in the Brownian limit, so I have made the choice that has slightly simpler algebra.

Consider the lower barrier at $i = 0$. The inflow to 1 comes from 1 itself (via the reflection at 0) with probability q, and from 2 with probability q. Thus

$$\Pi_1 = q\Pi_1 + q\Pi_2, \text{ or } \Pi_2 = (p/q)\Pi_1.$$

Near the upper barrier, similarly

$$\Pi_{m-1} = p\Pi_{m-1} + p\Pi_{m-2}, \text{ or } \Pi_{m-1} = (p/q)\Pi_{m-2}.$$

At all interior points (6.12) applies. Then we can inductively prove

$$\Pi_{i+1} = (p/q)\Pi_i \text{ for } i = 2, 3, \ldots (m-3).$$

For example,

$$\Pi_3 = (\Pi_2 - p\Pi_1)/q = (\Pi_2 - q\Pi_2)/q = (p/q)\Pi_2.$$

Then

$$\Pi_i = (p/q)^{i-1}\Pi_1 \text{ for } i = 1, 2, \ldots (m-1). \tag{6.15}$$

Finally, all the probabilities must add to unity. Therefore

$$1 = \sum_{i=1}^{m-1} \left(\frac{p}{q}\right)^{i-1} \Pi_1 = \frac{(p/q)^{m-1} - 1}{(p/q) - 1} \Pi_1. \tag{6.16}$$

If $p = q = \frac{1}{2}$, similar analysis leads to

$$\Pi_1 = \tfrac{1}{2}\Pi_1 + \tfrac{1}{2}\Pi_2, \text{ or } \Pi_1 = \Pi_2,$$

$$\Pi_{m-1} = \tfrac{1}{2}\Pi_{m-1} + \tfrac{1}{2}\Pi_{m-2}, \text{ or } \Pi_{m-2} = \Pi_{m-1},$$

and in the interior

$$\Pi_3 = 2\Pi_2 - \Pi_1 = \Pi_1$$

etc. Thus all the Π_i are equal, and

$$\Pi_i = 1/(m-1) \text{ for } i = 1, 2, \ldots (m-1). \tag{6.16'}$$

To find the Brownian motion limit, we use the expression (6.3) for powers of (q/p) and simplify. This yields the density function

$$\Pi(x) = \frac{2\mu}{\sigma^2} \frac{\exp(2x\mu/\sigma^2)}{\exp(2b\mu/\sigma^2) - \exp(2a\mu/\sigma^2)} \tag{6.17}$$

for $\mu \neq 0$, and

THE ART OF SMOOTH PASTING 61

$$\Pi(x) = 1/(b - a) \qquad (6.17')$$

when $\mu = 0$. As one would expect, the density is increasing with x when the trend parameter μ is positive, decreasing when it is negative, and uniform when it is zero.

The case of one reflecting barrier can be found by the usual limiting procedure. For example, fix a and let $b \to \infty$. If $\mu > 0$, then in (6.17) we have $\Pi(x) \to 0$ for all x; the distribution is degenerate and all the probability can be considered concentrated at ∞. This is intuitive as the long-run behavior of a motion with positive trend and no upper barrier. If $\mu = 0$ we again get degeneracy in (6.17'). If $\mu < 0$, then (6.17) gives a negative exponential density function away from the barrier:

$$\Pi(x) = (-2\mu/\sigma^2)\exp[2(x - a)\mu/\sigma^2]. \qquad (6.18)$$

Having found the density $\Pi(x)$, we can find the (undiscounted) long-run average of x, or indeed of any function $f(x)$, as

$$\int_a^b f(x)\Pi(x)dx.$$

For example, Bentolila and Bertola (1990) compute long-run average employment by a firm whose production function undergoes multiplicative Brownian shocks, and there are hiring and firing costs. We can also find the stationary distribution of any function of x. For example, in the target zone model of Section 3.7, Svensson (1991) finds the average of the exchange rate when the fundamental is regulated between the appropriate reflecting barriers.

Now suppose the reflecting barriers arise because there are linear costs of adjusting x and optimal barrier control is exercised as in Section 4.6. We can use (6.15-16) to calculate the (undiscounted) long run average costs of such control. Consider the lower barrier. Reflections occur on a down move from 1, that is a proportion $q\Pi_1$ of the time. At each such occasion, costs $c_l \Delta h$ are incurred to reflect the particle to location 1, where Δh is the distance between two adjacent points and c_l is the cost per unit distance at the lower barrier. Each unit period is of length Δt. Therefore the average cost per unit time is

$$\frac{q\Pi_1 c_l \Delta h}{\Delta t} = \frac{c_l(p - q)}{(p/q)^m - 1} \frac{\Delta h}{\Delta t}$$

$$= \frac{c_l \mu \Delta h/\sigma^2}{\exp(2(b-a)\mu/\sigma^2) - 1} \frac{\Delta h}{\Delta t}$$

$$= \frac{c_l \mu}{\exp(2(b-a)\mu/\sigma^2) - 1}. \tag{6.19}$$

Similarly, writing c_u for the cost per unit distance at the upper barrier, we get the (undiscounted) long-run average cost of action there

$$c_u \mu / [1 - \exp(-2(b-a)\mu/\sigma^2)].$$

When $\mu = 0$, these become $c_l \sigma^2/[2(b-a)]$ and $c_u \sigma^2/[2(b-a)]$, respectively.

For rigorous derivation of these expressions, see Harrison (1985, pp. 90–92).

Resetting

For algebraic simplicity I shall offer a particularly simple symmetric example of a calculation of long-run distribution with resetting. Consider a discrete random walk with equal probabilities of up and down moves ($p = q = \frac{1}{2}$) within barries at $\pm m$. Suppose a particle that moves down from $-(m-1)$ to hit the barrier at $-m$ is instantly moved to 0 and the motion restarted. Similarly at the upper barrier. As with reflecting barriers, I am assuming that the actual barrier locations are never occupied.

Consider the barrier location m. The location $(m-1)$ gets inflow only when an up move occurs from $(m-2)$. Therefore the stationary probabilities satisfy $\Pi_{m-1} = \frac{1}{2}\Pi_{m-2}$. Write $\Pi_{m-1} = A$, then $\Pi_{m-2} = 2A$. Next

$$\Pi_{m-2} = \tfrac{1}{2}(\Pi_{m-1} + \Pi_{m-3}), \quad \text{so}$$
$$\Pi_{m-3} = 2\Pi_{m-2} - \Pi_{m-1} = 4A - A = 3A.$$

Similarly

$$\Pi_{m-4} = 2\Pi_{m-3} - \Pi_{m-2} = 6A - 2A = 4A, \ldots$$
$$\Pi_0 = 2\Pi_1 - \Pi_2 = 2(m-1)A - (m-2)A = mA.$$

By symmetry we should have $\Pi_{-(m-1)} = A$, and proceeding to the right reach $\Pi_0 = mA$ again. Finally, the resulting Π_0 must be reconciled with the inflows to 0, which occur with the usual up and down steps from -1 and 1, and because of resettings. Thus

$$\Pi_0 = \tfrac{1}{2}(\Pi_{-1} + \Pi_1) + \tfrac{1}{2}(\Pi_{-(m-1)} + \Pi_{m-1})$$
$$= \tfrac{1}{2} 2(m-1)A + \tfrac{1}{2} 2A = mA$$

as required.

It remains to compute A by imposing the adding-up requirement. Using symmetry, we can write this as

$$1 = \Pi_0 + 2(\Pi_1 + \Pi_2 + \ldots + \Pi_{m-1})$$
$$= mA + 2(A + 2A + \ldots + 2(m-1)A)$$
$$= mA + 2\tfrac{1}{2}(m-1)mA$$
$$= m^2 A. \qquad (6.20)$$

For the discrete random walk, the stationary probabilities increase linearly toward the origin. Therefore in the Brownian motion limit, the density function becomes triangular. With the barriers at $\pm b = \pm m \Delta h$, we have

$$\Pi(x) = (b - |x|)/b^2. \qquad (6.21)$$

With this density function, we can compute the (undiscounted) long-run average of any function of x; for example in the menu cost model of section 4.5 we can calculate the average of the flow cost x^2.

We can also calculate the (undiscounted) long-run average cost of resetting. Each unit time interval is of length Δt. A proportion $\tfrac{1}{2}(\Pi_{m-1} + \Pi_{-(m-1)}) = A$ of these lead to resettings. If each resetting costs k, the average cost per unit time is

$$Ak/\Delta t = k/[\Delta t m^2] = k(\Delta h)^2/[\Delta t b^2].$$

In the Brownian limit, this becomes

$$k\sigma^2/b^2. \qquad (6.22)$$

For an application to the menu cost model, see Dixit (1992 b).

For a much more general model of long-run distributions when the underlying Brownian motion has a non-zero drift, and the resetting from upper and lower barriers can be to different internal points, see Bertola and Caballero (1990).

6.3. Dynamics of Brownian paths: Kolmogorov equations

Having examined some particular features of the short-run and long-run behavior of Brownian motion subject to barriers, I shall now derive a pair of equations governing more general dynamics, and briefly indicate how they can be solved.

Consider two points in time, t_0 and t_1, with $t_0 < t_1$. Let x_0 and x_1 be any two points in the range of variation of our Brownian particle. Denote by $\Pi(x_0, t_0, x_1, t_1)$ the conditional probability density of finding the particle at location x_1 at time t_1, given that it started at location x_0 at time t_0. For Brownian motion the initial state x_0 captures the entire effect of the history before t_0. The separate times t_0 and t_1 do not matter only, 'elapsed time' $(t_1 - t_0)$ matters. But for conceptual clarity we will carry the two arguments separately.

The equations I shall derive are named after Kolmogorov, an early researcher in this field. They pertain to the effect of varying (x_1, t_1) for given (x_0, t_0) — the Forward Equation — and vice versa — the Backward Equation.

I shall use again the discrete random walk approximation to Brownian motion. As before, Δh denotes the step size and Δt the length of the unit time period. To save notation, however, I shall not introduce the discrete location labels i, instead tacitly assuming that the points labelled (x_1, t_1) etc. lie on the discrete grid.

The Forward Equation
Consider the point $(x_1, t_1 + \Delta t)$. This could have arisen in either of two ways: $(x_1 - \Delta h, t_1)$ followed by an up move, or $(x_1 + \Delta h, t_1)$ followed by a down move. Therefore

$$\Pi(x_1, t_1 + \Delta t) = p\Pi(x_1 - \Delta h, t_1) + q\Pi(x_1 + \Delta h, t_1),$$

where for brevity I have left out the arguments (x_0, t_0) that are held fixed throughout this calculation.

Using the expressions (1.5') for p and q, and expanding the various values of the Π function in Taylor series, we get

$$\Pi(x_1, t_1) + \Pi_t(x_1, t_1)\Delta t + o(\Delta t)$$
$$= \tfrac{1}{2}\left[1 + \frac{\mu}{\sigma^2}\Delta h\right][\Pi(x_1, t_1) - \Pi_x(x_1, t_1)\Delta h$$
$$+ \tfrac{1}{2}\Pi_{xx}(x_1, t_1)(\Delta h)^2 + o(\Delta h)^2]$$
$$+ \tfrac{1}{2}\left[1 - \frac{\mu}{\sigma^2}\Delta h\right][\Pi(x_1, t_1) + \Pi_x(x_1, t_1)\Delta h$$
$$+ \tfrac{1}{2}\Pi_{xx}(x_1, t_1)(\Delta h)^2 + o(\Delta h)^2]$$

$$= \Pi(x_1, t_1) - \frac{\mu}{\sigma^2}(\Delta h)^2 \Pi_x(x_1, t_1) + \tfrac{1}{2}(\Delta h)^2 \Pi_{xx}(x_1, t_1)$$
$$+ o(\Delta h)^2,$$

where $o(\Delta t)$ denotes terms that tend to zero faster than Δt, and similarly $o(\Delta h)^2$. Finally, recalling the relation (1.4) between Δh and Δt, we get

$$\tfrac{1}{2}\sigma^2 \Pi_{xx}(x_1, t_1) - \mu \Pi_x(x_1, t_1) = \Pi_t(x_1, t_1). \tag{6.23}$$

This is a *partial* differential equation, with two independent variables (x_1, t_1), and one dependent variable Π. Solution of such equations is far more difficult than that of ordinary differential equations with just one independent variable. I shall merely indicate some methods that are useful in the particular context of (6.23). Interested readers will find a good recent exposition of the theory of partial differential equations in Guenther and Lee (1988); their Chapters 5 and 9 deal with 'parabolic' equations that are general versions of (6.23) above.

My derivation has followed Cox and Miller (1965, p. 208), who then use the special structure of the equation to solve it by Laplace transform methods (ibid, pp. 210–213). Here I shall outline a different method, namely separation of variables. First try a solution of the form $\Pi(x_1, t_1) = X(x_1)T(t_1)$. Substituting in (6.23) and regrouping terms, we have

$$[\tfrac{1}{2}\sigma^2 X''(x_1) - \mu X'(x_1)]/X(x_1) = T'(t_1)/T(t_1).$$

Since the left hand side is a function of x_1 alone and the right hand side a function of t_1 alone, each must be a constant, say λ. Then

$$\tfrac{1}{2}\sigma^2 X''(x_1) - \mu X'(x_1) - \lambda X(x_1) = 0$$

and

$$T'(t_1) = \lambda T(t_1).$$

These equations are easy to solve. The general solutions are

$$X(x_1) = A(\lambda) \exp[-\alpha(\lambda)x_1] + B(\lambda) \exp[\beta(\lambda)x_1]$$

and

$$T(t_1) = C(\lambda) \exp(\lambda t_1),$$

where $A(\lambda)$, $B(\lambda)$ and $C(\lambda)$ are constants to be determined, and $-\alpha(\lambda)$, $\beta(\lambda)$ are roots of the familiar quadratic

$$\phi(\xi) \equiv \lambda + \mu\xi - \tfrac{1}{2}\sigma^2\xi^2 = 0.$$

Note how we have allowed all these entities to depend on the free parameter λ that was introduced above. Not all values of λ are permissible; conditions of convergence etc. impose some limits on the range of λ, say $\lambda \in \Lambda$. The general solution of (6.23) can then be expressed as a linear combination of separable solutions of this kind, taken over the permissible range:

$$\Pi(x_1, t_1) = \int_\Lambda \{A(\lambda)\exp[-\alpha(\lambda)x_1]$$
$$+ B(\lambda)\exp[\beta(\lambda)x_1]\}\exp(\lambda t_1)d\lambda. \qquad (6.24)$$

I have subsumed $C(\lambda)$ into $A(\lambda)$ and $B(\lambda)$, but it remains to determine these functions. For that we have initial and boundary conditions.

The simplest initial condition is a given starting point (x_0, t_0). Then $\Pi(x_0, t_0)$ has all the density concentrated at x_0; it is a Dirac Delta Function. More generally and sometimes of more economic interest, we may be studying not a single particle but a large mass, for example firms or consumers, whose starting points differ for historical reasons. Then $\Pi(x_0, t_0)$ is this initial distribution. If the mass is large enough, the frequency distribution of this population replicates the probability distribution found in (6.24). The solution of the forward equation gives us the non-stochastic population dynamics of this mass of firms or consumers. For such an application to investment, see Bertola and Caballero (1991, Appendix B).

If the process is subject to barriers, there will be other conditions that must be met there. To derive these, we must modify the reasoning that led to (6.23). Consider an upper barrier at b. First suppose this is a reflecting barrier, and let the point x_1 be adjacent to the barrier in our discrete grid. The point $(x_1, t_1 + \Delta h)$ could arise from (x_1, t_1) after an up step and an instantaneous reflection, or from $(x_1 - \Delta h, t_1)$ after an up step. Thus

$$\Pi(x_1, t_1 + \Delta h) = p\Pi(x_1 - \Delta h, t_1) + p\Pi(x_1, t_1).$$

Expanding in Taylor series as before, we find

THE ART OF SMOOTH PASTING 67

$$\Pi(x_1, t_1) + \Pi_t(x_1, t_1)\Delta t + o(\Delta t) = \Pi(x_1, t_1)$$
$$+ \left[\frac{\mu}{\sigma^2}\Pi(x_1, t_1) - \tfrac{1}{2}\Pi_x(x_1, t_1)\right]\Delta h + o(\Delta h).$$

Now proceed to the Brownian limit. Note that $x_1 \to b$ as $\Delta h \to 0$. Also, since Δt goes to zero faster than Δh, we must have

$$\frac{\mu}{\sigma^2}\Pi(b, t_1) - \tfrac{1}{2}\Pi_x(b, t_1) = 0,$$

or

$$\Pi_x(b, t_1) = (2\mu/\sigma^2)\Pi(b, t_1) \tag{6.25}$$

for all $t_1 > t_0$. Next, if the barrier is absorbing, similar analysis gives

$$\Pi(b, t_1) = 0 \tag{6.26}$$

for all $t_1 > t_0$. For more details on such boundary conditions and their uses, see Cox and Miller (1965, pp. 219-225).

Long-run distribution
The only use of the Kolmogorov forward equation I shall discuss is for the long-run stationary distribution between two reflecting barriers at a and b. In such a distribution the influence of the initial condition should have died away, so the hidden arguments (x_0, t_0) in the function Π in (6.23) should be genuinely absent. Nor should the stationary distribution depend on the current time t_1. Then, replacing x_1 by x for notational simplicity, (6.23) reduces to the ordinary differential equation

$$\tfrac{1}{2}\sigma^2\Pi''(x) - \mu\Pi'(x) = 0. \tag{6.27}$$

If $\mu \neq 0$, one integration gives

$$\Pi'(x) = b\exp(2\mu x/\sigma^2),$$

where b is a constant of integration, and then another integration yields

$$\Pi(x) = A + B\exp(2\mu x/\sigma^2) \tag{6.28}$$

where A and B are constants. This has the same form as (6.14). If $\mu = 0$, we have $\Pi''(x) = 0$ and we can integrate twice to get

$$\Pi(x) = A + Bx \qquad (6.28')$$

where A and B are constants.

The conditions at the barriers become

$$\Pi'(a) = (2\mu/\sigma^2)\Pi(a), \quad \Pi'(b) = (2\mu/\sigma^2)\Pi(b). \qquad (6.29)$$

When $\mu = 0$, this gives $B = 0$ in (6.28'), so $\Pi(x)$ is uniform. Then the condition that the total probability should equal 1 leads to $\Pi(x) = 1/(b - a)$, same as (6.17') above. When $\mu \neq 0$, the barrier condition becomes $A = 0$ in (6.28), and the adding-up condition fixes B to yield the same exponential distribution as (6.17) above.

Bertola and Caballero (1991, Appendix B) carry out the much more difficult task of solving the forward equation for general finite time with reflecting barriers at a and b.

The Backward equation

Now fix (x_1, t_1) and consider the behavior of $\Pi(x_0, t_0)$. Starting at $(x_0, t_0 - \Delta t)$ leads to either $(x_0 + \Delta h, t_0)$ with probability p or $(x_0 - \Delta h, t_0)$ with probability q. Therefore

$$\Pi(x_0, t_0 - \Delta t) = p\Pi(x_0 + \Delta h, t_0) + q\Pi(x_0 - \Delta h, t_0).$$

Expanding and going to the limit as before, we have the Kolmogorov Backward Equation

$$\tfrac{1}{2}\sigma^2 \Pi_{xx}(x_0, t_0) + \mu \Pi_x(x_0, t_0) = -\Pi_t(x_0, t_0). \qquad (6.30)$$

This is very similar to the forward equation (6.23) except for a couple of crucial signs. One can also derive a mixed equation, for example focusing on x_0 and t_1 holding t_0 and x_1 constant. Thus

$$\tfrac{1}{2}\sigma^2 \Pi_{xx}(x_0, t_1) + \mu \Pi_x(x_0, t_1) = \Pi_t(x_0, t_1).$$

Sometimes this, rather than (6.30) above, is called the backward equation; see Cox and Miller (1965, pp. 209, 215) for both usages. Since only the 'elapsed time' $(t_1 - t_0)$ matters, the two forms differ only in the sign of the time derivative.

For more on the Kolmogorov equations, in addition to the above references to Cox and Miller (1965), see Bhattacharya and Waymire (1989, Chapter 5), and Karatzas and Shreve (1988, section 5.7).

References

Abel, Andrew B. (1983) 'Optimal investment under uncertainty,' *American Economic Review*, 73, 228-33.
Aitchison, J. and J. A. C. Brown (1957) *The Lognormal Distribution*, Cambridge, UK: Cambridge University Press.
Bartolini, Leonardo and Avinash Dixit (1991) 'Market valuation of illiquid debt and implications for conflict among creditors,' *IMF Staff Papers*, 38, 828-849.
Bentolila, Samuel and Giuseppe Bertola (1990) 'Firing costs and labor demand: How bad is Eurosclerosis?' *Reviews of Economic Studies*, 57, 381-402.
Bertola, Giuseppe and Ricardo Caballero (1990) 'Kinked adjustment costs and aggregate dynamics,' in *NBER Macroeconomics Annual 1990*, eds. Olivier Blanchard and Stanley Fischer, Cambridge, MA: MIT Press.
Bertola, Giuseppe and Ricardo Caballero (1991) 'Irreversibility and aggregate investment,' working paper, Princeton University.
Bhattacharya, Rabi N. and Waymire, Edward C. (1989) *Stochastic Processes with Applications*, New York: Wiley.
Constantinides, George M. and Scott, F. Richard (1978) 'Existence of optimal simple policies for discounted-cost inventory and cash management in continuous time, *Operations Research*, 26, 620-636.
Cox, David R. and H. D. Miller (1965) *The Theory of Stochastic Processes*, London: Chapman and Hall.
Cox, John C. and Stephen A. Ross (1976) 'The valuation of options for alternative stochastic processes,' *Journal of Financial Economics*, 3, 145-166.
Cox, John C., Stephen A. Ross and Mark Rubinstein (1979) 'Option pricing: A simplified approach,' *Journal of Financial Economics*, 7, 229-263.
Dixit, Avinash, (1991 a) 'Analytical approximations in models of hysteresis,' *Review of Economic Studies*, 58, 141-151.
Dixit, Avinash, (1991 b) 'A simplified treatment of the theory of optimal regulation of Brownian motion,' *Journal of Economic Dynamics and Control*, 15, 657-673.
Dixit, Avinash (1992 a) 'Investment and hysteresis,' *Journal of Economic Perspectives*, 6, 107-132.
Dixit, Avinash (1992 b) 'Space, time, and hysteresis,' working paper, Princeton University.
Dixit, Avinash and Robert Pindyck (1992) *Investment Under Uncertainty*, manuscript in preparation.
Dumas, Bernard (1991) 'Super contact and related optimality conditions,' *Journal of Economic Dynamics and Control*, 15, 675-85.
Feller, William (1968) *An Introduction to Probability Theory and Its Applications*, Volume I, Third Edition, New York: Wiley.
Feller, William (1971) *An Introduction to Probability Theory and Its Applications*, Volume II, Second Edition, New York: Wiley.
Fleming, W. H. and R. W. Rishel (1975) *Deterministic and Stochastic Optimal Control*, (Springer-Verlag, Berlin).
Gerlach, Stefan (1991) 'Adjustable pegs vs. single currencies: How valuable is the option to realign?' working paper, Brandeis University.
Guenther, Ronald B. and John W. Lee (1988) *Partial Differential Equations of Mathematical Physics and Integral Equations*, Englewood Cliffs, NJ: Prentice-Hall.
Harrison, J. Michael (1985) *Brownian Motion and Stochastic Flow Systems*, New York: Wiley.

Harrison, J. Michael and Michael I. Taksar (1983) 'Instantaneous control of Brownian motion,' *Mathematics of Operations Research*, **8**, 439-453.
Harrison, J. Michael, Thomas M. Sellke and Allison J. Taylor (1983) 'Impulse control of Brownian motion,' *Mathematics of Operations Research*, **8**, 454-466.
Karatzas, Ioannis and Steven E. Shreve (1988) *Brownian Motion and Stochastic Calculus*, (Springer-Verlag, Berlin).
Karlin, Samuel and Howard M. Taylor (1975) *A First Course in Stochastic Processes*, Second Edition, New York: Academic Press.
Karlin, Samuel and Howard M. Taylor (1981) *A Second Course in Stochastic Processes*, New York: Academic Press.
Krugman, Paul R. (1991) 'Target zones and exchange rate dynamics,' *Quarterly Journal of Economics*, **106**, 669-682.
Malliaris, Anastasios G. and William A. Brock (1983) *Stochastic Methods in Economics and Finance*, Amsterdam: North-Holland.
McDonald, Robert L. and Daniel R. Siegel (1985) 'Investment and the valuation of firms when there is an option to shut down,' *International Economic Review*, **26**, 331-349.
McDonald, Robert L. and Daniel R. Siegel (1986) 'The value of waiting to invest,' *Quarterly Journal of Economics*, **101**, 707-727.
Pindyck, Robert S. (1991) 'Irreversibility, uncertainty, and investment,' *Journal of Economic Literature,* **29**, 1110-1148.
Svensson, Lars E. O. (1991) 'Target zones and interest rate variability,' *Journal of International Economics*, **31**, 27-54.
Svensson, Lars E. O. (1992) 'Recent research on exchange rate target zones: An interpretation,' *Journal of Economic Perspectives*, **6**, 119-144.
Weatherburn, C. E. (1946) *A First Course in Mathematical Statistics*, Cambridge, UK: Cambridge University Press.

Index

Abel, Andrew B. 39
Aitchison, J. 8, 10
Barriers 22
  absorbing 22-23, 33, 67
  reflecting 22-23, 34, 59, 66, 67
Bartolini, Leonardo 49
Bentolila, Samuel 58, 61
Bertola, Giuseppe 58, 61, 63, 66, 68
Bhattacharya, Rabi N. 4, 57, 68
Brock, William A. 27, 34
Brown, J.A.C. 8, 10
Brownian motion 1-4
  absolute 7
  geometric 6, 8

Caballero, Ricardo 63, 66, 68
Competitive industry 45-46
Constantinides, George M. 43
Control 33
  barrier 34, 42
  continuous 38-39
  impulse 33, 39-41
  instantaneous 34
  optimal 33
Cox, D.R. 4, 53, 57, 65, 67, 68

Differential equation for present value 15-17
  case of geometric Brownian motion 19-21
  solution with barriers 24-25
Diffusion process 8, 21
Discounted present value 9
  for exponential 11
  for polynomial 12-13
  for powers of geometric Brownian motion 13
Dixit, Avinash 38, 41, 45, 47, 49, 58, 63
Dumas, Bernard 42

Exchange rate target zones 28-30

Expected time to barrier 54-57

Feller, William 3, 53, 57
Flemming, W.H. 34
Fundamental quadratic 11
  for geometric Brownian motion 14

Gambler's ruin 52-53
Gerlach, Stefan 49
Guenther, Ronald B. 65

Harrison, J. Michael 4, 6, 34, 43, 57, 62

Irreversible investment 37-38
Ito calculus 1, 5, 8
Ito process 8
Ito's Lemma 4-6

Jensen's inequality 6

Karatzas, Ioannis 34, 68
Karlin, Samuel 4, 6, 57
Kolmogorov equations 64
  backward 68
  forward 64-66
Krugman, Paul R. 30

Lee, John W. 65
Long run distribution 58
  between reflecting barriers 59-60, 67-68
  with resetting 62-63

Malliaris, Anastasios G. 27, 34
McDonald, Robert L. 31, 37
Mean-reversion 9, 47
Menu costs 41, 57
Miller, H.D. 4, 53, 57, 65, 67, 68

Pindyck, Robert S. 38, 47, 50
Price ceiling 27-28
Probability of reaching barrier 53-54

INDEX

Random walk 2, 52
Regulation 33
Resetting 26, 33, 62
  optimal 39, 43-44
Richard, Scott F. 43
Rishel, R.W. 34

Sellke, Thomas M. 34, 43
Shreve, Steven E. 34, 68
Siegel, Daniel R. 31, 37
Smooth pasting condition 27, 29, 36-37, 40, 41, 42, 43-44, 48, 51, 57
Stopping 26
  optimal 34-36

Super contact condition 42, 45, 46
Svensson, Lars E.O. 30, 61

Taskar, Michael I. 34
Taylor, Allison J. 34, 43
Taylor, Howard M. 4, 6, 57

Value matching condition 26, 40, 41, 43-44, 48, 51

Waymire, Edward C. 4, 57, 68
Weatherburn, C.E. 10
Wiener process 1

FUNDAMENTALS OF PURE AND APPLIED ECONOMICS
SECTIONS AND EDITORS

BALANCE OF PAYMENTS AND INTERNATIONAL FINANCE
W. Branson, Princeton University
DISTRIBUTION
A. Atkinson, London School of Economics
ECONOMIC DEVELOPMENT STUDIES
S. Chakravarty, Delhi School of Economics
ECONOMIC HISTORY
P. David, Stanford University, and M. Lévy-Leboyer, Université Paris X
ECONOMIC SYSTEMS
J.M. Montias, Yale University
ECONOMICS OF HEALTH, EDUCATION, POVERTY AND CRIME
V. Fuchs, Stanford University
ECONOMICS OF THE HOUSEHOLD AND INDIVIDUAL BEHAVIOR
J. Muellbauer, University of Oxford
ECONOMICS OF TECHNOLOGICAL CHANGE
F.M. Scherer, Harvard University
EVOLUTION OF ECONOMIC STRUCTURES, LONG-TERM MODELS, PLANNING POLICY, INTERNATIONAL ECONOMIC STRUCTURES
W. Michalski, O.E.C.D., Paris
EXPERIMENTAL ECONOMICS
C. Plott, California Institute of Technology
GOVERNMENT OWNERSHIP AND REGULATION OF ECONOMIC ACTIVITY
E. Bailey, Carnegie-Mellon University, USA
INTERNATIONAL ECONOMIC ISSUES
B. Balassa, The World Bank
INTERNATIONAL TRADE
M. Kemp, University of New South Wales
LABOR AND ECONOMICS
F. Welch, University of California, Los Angeles, and J. Smith, The Rand Corporation
MACROECONOMIC THEORY
J. Grandmont, CEPREMAP, Paris

MARXIAN ECONOMICS
J. Roemer, University of California, Davis
NATURAL RESOURCES AND ENVIRONMENTAL ECONOMICS
C. Henry, Ecole Polytechnique, Paris
ORGANIZATION THEORY AND ALLOCATION PROCESSES
A. Postlewaite, University of Pennsylvania
POLITICAL SCIENCE AND ECONOMICS
J. Ferejohn, Stanford University
PROGRAMMING METHODS IN ECONOMICS
M. Balinski, Ecole Polytechnique, Paris
PUBLIC EXPENDITURES
P. Dasgupta, University of Cambridge
REGIONAL AND URBAN ECONOMICS
R. Arnott, Boston College, Massachusetts
SOCIAL CHOICE THEORY
A. Sen, Harvard University
STOCHASTIC METHODS IN ECONOMIC ANALYSIS
Editor to be announced
TAXES
R. Guesnerie, Ecole des Hautes Etudes en Sciences Sociales, Paris
THEORY OF THE FIRM AND INDUSTRIAL ORGANIZATION
A. Jacquemin, Université Catholique de Louvain

FUNDAMENTALS OF PURE AND APPLIED ECONOMICS

PUBLISHED TITLES

Volume 1 (International Trade Section)
GAME THEORY IN INTERNATIONAL ECONOMICS
by John McMillan

Volume 2 (Marxian Economics Section)
MONEY, ACCUMULATION AND CRISIS
by Duncan K. Foley

Volume 3 (Theory of the Firm and Industrial Organization Section)
DYNAMIC MODELS OF OLIGOPOLY
by Drew Fudenberg and Jean Tirole

Volume 4 (Marxian Economics Section)
VALUE, EXPLOITATION AND CLASS
by John E. Roemer

Volume 5 (Regional and Urban Economics Section)
LOCATION THEORY
by Jean Jaskold Gabszewicz and Jacques-François Thisse, Masahisa Fujita, and Urs Schweizer

Volume 6 (Political Science and Economics Section)
MODELS OF IMPERFECT INFORMATION IN POLITICS
by Randall L. Calvert

Volume 7 (Marxian Economics Section)
CAPITALIST IMPERIALISM, CRISIS AND THE STATE
by John Willoughby

Volume 8 (Marxian Economics Section)
MARXISM AND "REALLY EXISTING SOCIALISM"
by Alec Nove

Volume 9 (Economic Systems Section)
THE NONPROFIT ENTERPRISE IN MARKET ECONOMIES
by Estelle James and Susan Rose-Ackerman

Volume 10 (Regional and Urban Economics Section)
URBAN PUBLIC FINANCE
by David E. Wildasin

Volume 11 (Regional and Urban Economics Section)
URBAN DYNAMICS AND URBAN EXTERNALITIES
by Takahiro Miyao and Yoshitsugu Kanemoto

Volume 12 (Marxian Economics Section)
DEVELOPMENT AND MODES OF PRODUCTION IN MARXIAN
ECONOMICS: A CRITICAL EVALUATION
by Alan Richards

Volume 13 (Economics of Technological Change Section)
TECHNOLOGICAL CHANGE AND PRODUCTIVITY GROWTH
by Albert N. Link

Volume 14 (Economic Systems Section)
ECONOMICS OF COOPERATION AND THE LABOR-MANAGED
ECONOMY
by John P. Bonin and Louis Putterman

Volume 15 (International Trade Section)
UNCERTAINTY AND THE THEORY OF INTERNATIONAL TRADE
by Earl L. Grinols

Volume 16 (Theory of the Firm and Industrial Organization Section)
THE CORPORATION: GROWTH, DIVERSIFICATION AND MERGERS
by Dennis C. Mueller

Volume 17 (Economics of Technological Change Section)
MARKET STRUCTURE AND TECHNOLOGICAL CHANGE
by William L. Baldwin and John T. Scottt

Volume 18 (Social Choice Theory Section)
INTERPROFILE CONDITIONS AND IMPOSSIBILITY
by Peter C. Fishburn

Volume 19 (Macroeconomic Theory Section)
WAGE AND EMPLOYMENT PATTERNS IN LABOR CONTRACTS:
MICROFOUNDATIONS AND MACROECONOMIC IMPLICATIONS
by Russell W. Cooper

Volume 20 (Government Ownership and Regulation of Economic Activity Section)
DESIGNING REGULATORY POLICY WITH LIMITED INFORMATION
by David Besanko and David E.M. Sappington

Volume 21 (Economics of Technological Change Section)
THE ROLE OF DEMAND AND SUPPLY IN THE GENERATION AND
DIFFUSION OF TECHNICAL CHANGE
by Colin G. Thirtle and Vernon W. Ruttan

Volume 22 (Regional and Urban Economics Section)
SYSTEMS OF CITIES AND FACILITY LOCATION
by Pierre Hansen, Martine Labbé, Dominique Peeters and Jacques-François
Thisse, and J. Vernon Henderson

Volume 23 (International Trade Section)
DISEQUILIBRIUM TRADE THEORIES
by Motoshige Itoh and Takashi Negishi

Volume 24 (Balance of Payments and International Finance Section)
THE EMPIRICAL EVIDENCE ON THE EFFICIENCY OF FORWARD AND FUTURES FOREIGN EXCHANGE MARKETS
by Robert J. Hodrick

Volume 25 (Economic Systems Section)
THE COMPARATIVE ECONOMICS OF RESEARCH DEVELOPMENT AND INNOVATION IN EAST AND WEST: A SURVEY
by Philip Hanson and Keith Pavitt

Volume 26 (Regional and Urban Economics Section)
MODELING IN URBAN AND REGIONAL ECONOMICS
by Alex Anas

Volume 27 (Economic Systems Section)
FOREIGN TRADE IN THE CENTRALLY PLANNED ECONOMY
by Thomas A. Wolf

Volume 28 (Theory of the Firm and Industrial Organization Section)
MARKET STRUCTURE AND PERFORMANCE - THE EMPIRICAL RESEARCH
by John S. Cubbin

Volume 29 (Economic Development Studies Section)
STABILIZATION AND GROWTH IN DEVELOPING COUNTRIES: A STRUCTURALIST APPROACH
by Lance Taylor

Volume 30 (Economics of Technological Change Section)
THE ECONOMICS OF THE PATENT SYSTEM
by Erich Kaufer

Volume 31 (Regional and Urban Economics Section)
THE ECONOMICS OF HOUSING MARKETS
by Richard F. Muth and Allen C. Goodman

Volume 32 (International Trade Section)
THE POLITICAL ECONOMY OF PROTECTION
by Arye L. Hillman

Volume 33 (Natural Resources and Environmental Economics Section)
NON-RENEWABLE RESOURCES EXTRACTION PROGRAMS AND MARKETS
by John M. Hartwick

Volume 34 (Government Ownership and Regulation of Economic Activity Section)
A PRIMER ON ENVIRONMENTAL POLICY DESIGN
by Robert W. Hahn

Volume 35 (Economics of Technological Change Section)
TWENTY-FIVE CENTURIES OF TECHNOLOGICAL CHANGE
by Joel Mokyr

Volume 36 (Government Ownership and Regulation of Economic Activity Section)
PUBLIC ENTERPRISE IN MONOPOLISTIC AND OLIGOPOLISTIC INDUSTRIES
by Ingo Vogelsang

Volume 37 (Economic Development Studies Section)
AGRARIAN STRUCTURE AND ECONOMIC UNDERDEVELOPMENT
by Kaushik Basu

Volume 38 (Macroeconomic Theory Section)
MACROECONOMIC POLICY, CREDIBILITY AND POLITICS
by Torsten Persson and Guido Tabellini

Volume 39 (Economic History Section)
TYPOLOGY OF INDUSTRIALIZATION PROCESSES IN THE NINETEENTH CENTURY
by Sidney Pollard

Volume 40 (Marxian Economics Section)
THE STATE AND THE ECONOMY UNDER CAPITALISM
by Adam Przeworski

Volume 41 (Theory of the Firm and Industrial Organization Section)
BARRIERS TO ENTRY AND STRATEGIC COMPETITION
by Paul Geroski, Richard J. Gilbert, and Alexis Jacquemin

Volume 42 (Macroeconomic Theory Section)
REDUCED FORMS OF RATIONAL EXPECTATIONS MODELS
by Laurence Broze, Christian Gouriéroux, and Ariane Szafarz

Volume 43 (Theory of the Firm and Industrial Organization Section)
COMPETITION POLICY IN EUROPE AND NORTH AMERICA: ECONOMIC ISSUES AND INSTITUTIONS
by W.S. Comanor, K. George, A. Jacquemin, F. Jenny, E. Kantzenbach, J.A. Ordover, and L. Waverman

Volume 44 (Natural Resources and Environmental Economics Section)
MODELS OF THE OIL MARKET
by Jacques Crémer and Djavad Salehi-Isfahani

Volume 45 (Political Science and Economics Section)
MODELS OF MULTIPARTY ELECTORAL COMPETITION
by Kenneth A. Shepsle

Volume 46 (Political Science and Economics Section)
SIGNALING GAMES IN POLITICAL SCIENCE
by Jeffrey S. Banks

Volume 47 (Regional and Urban Economics Section)
URBAN AND REGIONAL ECONOMICS—MARXIST PERSPECTIVES
by Matthew Edel

Volume 48 (Economic History Section)
THE RISE OF THE AMERICAN BUSINESS CORPORATION
by Richard S. Tedlow

Volume 49 (Natural Resources and Environmental Economics Section)
LONG-TERM CONTROL OF EXHAUSTIBLE RESOURCES
by Pierre Lasserre

Volume 50 (Economic Systems Section)
CENTRAL PLANNING
by Paul G. Hare

Volume 51 (Regional and Urban Economics Section)
URBAN TRANSPORTATION ECONOMICS
by Kenneth A. Small

Volume 52 (Distribution Section)
EMPIRICAL STUDIES OF EARNINGS MOBILITY
by A.B. Atkinson, F. Bourguignon and C. Morrisson

Volume 53 (Organization Theory and Allocation Processes Section)
BAYESIAN IMPLEMENTATION
by Thomas R. Palfrey and Sanjay Srivastava

Volume 54 (International Trade Section)
THE WELFARE ECONOMICS OF INTERNATIONAL TRADE
by Murray C. Kemp and Henry Y. Wan, Jr.

Volume 55 (Stochastic Methods in Economic Analysis Section)
THE ART OF SMOOTH PASTING
by Avinash Dixit

Volume 56 (Distribution Section)
POVERTY COMPARISONS
by Martin Ravallion

Volume 57 (Economic Systems Section)
COMPARATIVE ECONOMICS
by John Michael Montias, Avner Ben-Ner and Egon Neuberger

Volume 58 (Political Science and Economics Section)
COMMITTEES, AGENDAS, AND VOTING
by Nicholas R. Miller

Volume 59 (Labor and Economics Section)
WELFARE AND THE WELL-BEING OF CHILDREN
by Janet M. Currie

Volume 60 (Labor and Economics Section)
EMPIRICAL METHODS FOR THE STUDY OF LABOR FORCE DYNAMICS
by Kenneth I. Wolpin

ISSN: 0191-1708